POULTRY LIGHTING
the theory and practice

Peter Lewis & Trevor Morris

NORTHCOT

Published by **Northcot**
Cowdown Lane
Goodworth Clatford
Andover
Hampshire SP11 7HG
United Kingdom

Tel: +44 1264 350556
Email: peter.lewis@dsl.pipex.com
Website: www.poultrylighting.co.uk

© Northcot 2006. All rights reserved. No part of this publication may be reproduced in any form or by any means, electronically, mechanically, by photocopying, recording or otherwise, without the prior permission of the copyright owners. Applications for copyright holder's permission to reproduce any part of this publication should be addressed to the publisher.

ISBN   0-9552104-0-2
       978-0-9552104-0-2

Printed and bound in the UK by The Cromwell Press, Trowbridge, Wiltshire BA14 0XB

The authors

Peter Lewis gained a National Diploma in Poultry Husbandry at Harper Adams Agricultural College, UK in 1964 and was awarded a PhD for his studies of interrupted lighting for laying hens at the University of Bristol in 1987. He has been a technical advisor for three primary poultry breeding companies, lectured in poultry husbandry and been involved in commercial broiler management. He has been secretary of the UK Branch of WPSA, Assistant Editor of World's Poultry Science Journal and is currently a director of British Poultry Science. Peter has held research fellowships at the Universities of Bristol and Reading, and is now a Research Fellow at the University of KwaZulu-Natal in South Africa researching the responses of broiler breeders to lighting. Peter is also an Adjunct Professor at the University of Guelph in Canada. He has produced more than 80 papers in scientific journals, most of which have been concerned with lighting, and has presented papers at various world congresses and regional conferences.

Trevor Morris read Agriculture at the University of Reading and then, after postgraduate training in Poultry Husbandry, joined the staff of the Department of Agriculture at Reading. He retired from the same Department 43 years later, having devoted his time to teaching and research in poultry science and collecting PhD and DSc degrees along the way. He is now an Emeritus Professor of the University of Reading. He has won numerous scientific and industrial awards for his research and served as President of the UK Branch of the WPSA and Senior Vice-President of the world association. His publications include books on energy requirements of poultry, poultry immunology, experimental design in the animal sciences and more than 200 refereed papers in scientific journals, about half of which deal with some aspect of photoperiodism.

In 1995, the authors, together with Dr Graham Perry, were awarded the Tom Newman Memorial medal for their contributions to our knowledge of poultry lighting.

Acknowledgements

Steve Leeson for the inspiration to even think about publishing the book ourselves.
Tom Siopes for the photograph used in Figure 2.2.
Sue Gordon for information on lighting for broilers.
Cliff Nixey and Nick French for invaluable discussion on the practicalities of lighting turkeys.
Andre Engler for information on lighting for Muscovy ducks.
Peter Cherry for editing the section on the practical lighting of ducks.
Cherry Valley for management information on lighting for Pekin ducks.
Sources of material used in figures and tables are acknowledged in the respective captions.

Contents

	page
Chapter 1. LIGHT	
History, physics and luminous efficiencies	1
Glossary	2
Chapter 2. PHYSIOLOGY AND MECHANISMS	
Avian vision and comparative luminous efficiencies	7
Photoreception and responses to illumination	9
Penetrability to, and sensitivity and intensity thresholds at, the hypothalamus of different wavelengths of light	11
Photoinducible phase	13
Phase reversal	14
Photoperiodic time measurement	14
Carry-over effect	15
Photorefractoriness	16
Chapter 3. PHOTOPERIOD: Conventional programmes	
Growing pullets: constant photoperiods	23
Growing pullets: changing photoperiods	25
Laying hens: constant photoperiods	28
Laying hens: changing photoperiods	30
Broiler breeders: constant photoperiods	32
Broiler breeders: changing photoperiods	34
Broilers	38
Growing turkeys	40
Breeding turkeys	42
Waterfowl	44
Muscovy (Barbarie)	44
Geese	45
Ducks	46
Dawn and dusk	49
Chapter 4. PHOTOPERIOD: Unconventional programmes	
Intermittent programmes	57
Growing pullets	59
Laying hens	59
Broiler breeders	61
Broilers	62
Breeding turkeys	63
Growing turkeys	64
Ahemeral programmes	65
Growing pullets	65
Laying hens	65
Broiler breeders	69
Breeding turkeys	70
Continuous illumination	70
Broilers	70
Growing turkeys	71
Cockerels and laying hens	71
Continuous darkness	72
Broilers	72
Growing pullets and laying hens	72

	page

Chapter 5. ILLUMINANCE (Light intensity)
Measurement and units of illuminance — 79
Growing pullets — 81
Laying hens — 82
Broiler breeders — 85
Broilers — 86
Breeding turkeys — 87
Growing turkeys — 89
Waterfowl — 89
  Geese — 89
  Ducks — 90

Chapter 6. WAVELENGTH (Colour)
Growth — 95
Male reproduction — 97
Female reproduction — 97
Behaviour, preference, stress and mortality — 98

Chapter 7. PATHOLOGICAL EFFECTS OF LIGHTING
Glossary — 103
Lighting extremes for chicken and turkeys — 104
  Constant illumination — 104
  Continuous darkness — 106
Illuminance and wavelength — 107
  Illuminance, eye pathology and stress — 107
  Wavelength, eye pathology and stress — 107
Ovarian carcinoma in turkeys — 108

Chapter 8. LIGHT SOURCE
Light generation and light sources — 111
  Light generation — 111
  Lamp characteristics — 112
  Perception of fluorescent light flicker — 113
  Fluorescent anti-rachitic properties — 114
Production responses to light source — 115
  Broilers — 115
  Growing pullets — 115
  Laying hens and broiler breeders — 115
  Breeding turkeys — 116
  Growing turkeys — 117
  Geese — 117
General conclusions — 117

Chapter 9. LIGHTING FOR GROWING PULLETS AND LAYING HENS
Rearing period — 121
  Daylength — 121
  Light intensity — 125
  Light colour and source — 125
Laying period — 125
  Daylength — 126
  Light intensity — 129
  Light colour and source — 130
  Intermittent lighting — 132
  Ahemeral lighting — 135

	page
Chapter 10. LIGHTING FOR BROILER BREEDERS	
Rearing period	137
Daylength	137
Light intensity	139
Light colour and source	139
Males	140
Laying period	140
Daylength	140
Light intensity	143
Light colour and source	144
Chapter 11. LIGHTING FOR BROILERS	
Daylength	145
Intermittent lighting	146
Light intensity	147
Light colour and source	147
Chapter 12. LIGHTING FOR BREEDING TURKEYS	
Females	149
Rearing daylength	149
Rearing light intensity	151
Photostimulation age	151
Laying daylength	152
Laying light intensity	152
Males	152
Daylength	153
Light intensity	153
Light colour and source	153
Recycling	153
Ovarian carcinoma	154
Polyovarian Follicle and Polycystic Ovarian Follicle Syndrome	154
Chapter 13. LIGHTING FOR GROWING TURKEYS	
Daylength	155
Intermittent lighting	158
Light intensity	159
Light colour	159
Light source	160
Chapter 14. LIGHTING FOR WATERFOWL	
Muscovy (Barbarie)	161
Daylength	161
Light intensity and colour	162
Geese	162
Daylength	162
Light intensity, colour and source	163
Ducks	164
Daylength for meat-type parents	164
Light intensity, colour and source for meat-type parents	164
Duckling	164
Daylength	164
Light intensity	164
Light intensity, colour and source	164
Index	165

1 LIGHT

This chapter deals with the physical aspects of light, including how the energy of a photon changes with wavelength and the relationship between radiometric and photometric units. It also includes a glossary of lighting terminology.

• History, physics and luminous efficiencies •

Light is the name given to visible electromagnetic radiation. It has also been considered to be a combination of radiation and the response to it (1). Light is but one part of a complex of physical phenomena called electromagnetic radiation. The nature of the phenomenon depends on its wavelength, beginning with cosmic rays at 10^{-18}m, and lengthening through gamma-rays, X-rays, ultraviolet, visible light (4.0 to 7.8 x 10^{-7} m or 400 to 780 nm), infrared, microwaves, radar, television and radio to power transmission at 6000 kilometres! Whereas different wavelengths of light result in the perception of different colour sensations (Table 1.1), reception of radiation across the range is perceived as white light. Changes in the wavelength of radiation can be achieved by using materials which absorb one wavelength and then give off another. For example, phosphors on the inside walls of low-pressure fluorescent lamps absorb the ultraviolet radiation generated by a mercury discharge and then emit visible light.

Two theories have been advanced to explain the nature of electromagnetic radiation. James Maxwell perfected his wave theory in 1862 (before the discovery of the electron), theorising that it consisted of waves propagated in electric and magnetic force fields, and that the different properties of the various types of electromagnetic radiation are the result of differences in their wavelength. He also proved that disturbances in the electric and magnetic conditions could be propagated through empty space with a definite velocity (speed of light = 2.998 x 10^8 m/s or 6.706 x 10^8 mph). The second theory, postulated in 1900 by the German physicist Max Planck, was that radiation consists of very small indivisible amounts of energy called 'quanta', and that different types of electromagnetic radiation consist of quanta of different energy content, with shorter wavelengths having higher amounts of energy (Table 1.1). Planck found that for a given wavelength the energy in a quantum (or photon) of visible light is the product of its frequency (Hz) and a constant (6.626 x 10^{-34} J.s). Frequency and wavelength therefore have a direct relationship, and so the energy content can be calculated from either characteristic:

$$(1)\ E = h.v \text{ or } (2)\ E = (h.c)/\lambda$$

where, E = quantum energy (joules), h (Planck's constant) = 6.626 x 10^{-34} (J.s), v = radiation frequency (Hz), c (velocity of the photon or speed of light in a vacuum) = 2.998 x 10^8 (m/s), and λ = radiation wavelength (m).

Table 1.1 *Quantum energy and human colour sensations at different wavelengths and frequencies of electromagnetic radiation.*

Wavelength (nm)	Frequency (Hz)	Quantum energy at mean wavelength (joules)	Human colour sensation
380 to 435	7.89 to 6.90 x 10^{14}	4.87 x 10^{-19}	violet
435 to 500	6.00 to 6.90 x 10^{14}	4.25 x 10^{-19}	blue
500 to 565	5.31 to 6.00 x 10^{14}	3.73 x 10^{-19}	green
565 to 600	5.00 to 5.31 x 10^{14}	3.41 x 10^{-19}	yellow
600 to 630	4.76 to 5.00 x 10^{14}	3.23 x 10^{-19}	orange
630 to 780	3.85 to 4.76 x 10^{14}	2.82 x 10^{-19}	red

Table 1.2 *Spectral luminous efficiencies (Vλ) for the photopic vision of humans (2) and poultry (3) at various wavelengths (λ).*

λ (nm)	V (λ) Human	V (λ) Poultry	λ (nm)	V (λ) Human	V (λ) Poultry
350	0.000	0.000	565	0.979	1.000
360	0.000	0.047	570	0.952	0.994
370	0.000	0.148	580	0.870	0.888
380	0.000	0.198	590	0.757	0.660
390	0.000	0.187	600	0.631	0.528
400	0.000	0.167	610	0.503	0.549
410	0.001	0.157	620	0.381	0.566
420	0.004	0.203	630	0.265	0.660
430	0.012	0.313	640	0.175	0.549
440	0.023	0.445	650	0.107	0.330
450	0.038	0.555	660	0.061	0.214
460	0.060	0.660	670	0.032	0.170
470	0.091	0.747	680	0.017	0.132
480	0.139	0.818	690	0.008	0.093
490	0.208	0.824	700	0.004	0.058
500	0.323	0.682	710	0.002	0.044
510	0.503	0.626	720	0.001	0.028
520	0.710	0.731	730	0.001	0.016
530	0.862	0.840	740	0.000	0.006
540	0.954	0.934	750	0.000	0.002
550	0.995	0.972	760	0.000	0.001
555	1.000	0.989	770	0.000	0.000
560	0.995	0.994	780	0.000	0.000

• Glossary •

Accommodation is the ability to sharply focus on objects at different distances. This is achieved by the ciliary muscles of the eye changing the focal length of the lens. For nearby objects the muscles contract to make the lens more convex, but they relax to make the lens flatter for objects further away (see Figure 7.1, page 103 for anatomy of the eye).

Acuity is the ability to see fine detail clearly. Visual acuity is 1/minimum angular size (the angle under which two objects can be seen separately), measured in minutes of arc, and in humans is determined by ascertaining the smallest letters that can be correctly identified at a given distance from a test chart. It will, however, be influenced by the light intensity and the time spent looking at an object.

Adaptation is the mechanism by which the eye alters its sensitivity as the light intensity or luminance changes. It is achieved by changing the diameter of the pupil and by altering the proportion of unstimulated rods and cones. Light changes the chemical composition of the photosensitive pigments when it enters the eye. In constant lighting conditions, the proportions of stimulated and unstimulated photopigment are about equal. The pigment is regenerated during darkness rendering it reactable to light, but cones regain sensitivity more rapidly than rods.

Ahemeral is the term applied to cycles of light and darkness that are longer or shorter than 24 h (Greek, *a hemera* - not a day).

Asymmetric fluorescent lamps do not have the same rate of discharge (e.g. 50/s) in each direction. As lamps age they can become asymmetric and their light output will appear as flicker to poultry.

Candela (cd) is the luminous intensity in a given direction of a light source emitting monochromatic radiation of frequency 540 x 10^{12} Hz with a radiant intensity in that direction of 1.46 x 10^{-3} (1/683) W/sr. It is a measure of the amount of light in a given direction, as opposed to luminous flux or power, which describes the total output from a source.

Circadian oscillators are biological functions that vary in intensity at an interval which is close, but not necessarily equal, to 24 hours (Latin, *circa* - about, *dies* - a day).

Circannual oscillators are biological functions that vary in intensity at an interval of about a year.

Civil twilight is the period when the sun is between 0 and 6° below the horizon, but still illuminates the sky.

Clux is a term that describes the illuminance perceived by poultry and is synonymous with gallilux. It differs from lux in that it is calculated using the luminous efficiencies of poultry (3) rather than humans (2).

Colour coding for lamps indicates the colour rendering index and colour temperature. It has three digits, the first being the colour rendering group and the next two indicate the colour temperature of the light (K). For example, an 827 lamp has a colour rendering index between 80 and 89 (group 8) and a colour temperature of 2700 K.

Colour rendering index (R_a) is a description of a light source's ability to reveal colours, and is based on 8 test colours established by CIE (Commission Internationale de l'Éclairage) in 1965. Sodium lamps have a very low rendering index (<25) because its yellow light makes colour differentiation almost impossible, in contrast, incandescent lamps have a 100 index and permit excellent colour perception.

Colour temperature is an indication of the colour appearance of white light, with warmer colours having lower temperatures. For example, warm white ≤ 3300 K, cool white 4000 K, and daylight ≥ 5000 K.

Cones are the retinal nerve endings that are concerned with colour vision, and named because of their shape. They are concentrated in the fovea of the eye and are capped by different oil droplets that permit different wavelengths of light to pass through to give different colour sensations. They produce sharp images because each cone generally has its own nerve connection, but are a lot fewer in number than rods and have lower light sensitivity. They progressively cease to function below an illuminance of about 4 lx.

Critical fusion frequency (CFF) is the frequency at which discontinuous light stimuli or flicker (e.g. from a fluorescent lamp) are seen as continuous. CFF decreases as the light intensity is reduced.

Dawn is the dark-light interface that an animal interprets as the beginning of its day.

Day is the part of the lighting cycle between dawn and dusk during which plasma melatonin concentrations are minimal, even though it may be only intermittently illuminated.

Dioptre (D) is the measure of the refractive power of the lens, and is the reciprocal of the focal length of the lens in metres.

Dusk is the light-dark interface that an animal interprets as the beginning of night.

Fluorescence is the emission of light by a material in response to stimulation by light of a different wavelength.

Flux is the basic unit of optical power, and is measured in watts (W) or lumens (lm).

Focal length is the distance from the point on the axis of the eye's lens where the rays of light cease to be parallel and converge or diverge to the centre of the lens.

Foot candle is the imperial measure of illuminance that is equal to 1 lm/ft² or 10.76 lx.

Frequency (v) or cycles per second is calculated from the energy in electron volts and Planck's constant (h) using the equation:

$$v = eV/h$$

where v = frequency (Hz), eV = electron volts (joules), and h = 6.626 x 10^{-34} (J.s).

Gallilux is a term that describes the illuminance perceived by poultry and is synonymous with clux. It differs from lux in that it is calculated using the luminous efficiencies of poultry (3) rather than humans (2).

Hue is the colour sensation created by the light – orange, blue, green etc.

Illuminance (E_v) is the luminous flux density on a surface, and is measured in lumens per square metre (lm/m²) or lux (lx). It is synonymous with light intensity or lighting level and is the photometric equivalent of irradiance. For a monochromatic light source it is calculated as:

$$E_v = W \cdot V_\lambda \cdot 683 \cdot (1/4\pi) \cdot (1/d^2)$$

where W = radiant flux in watts (W), V_λ = the spectral luminous efficiency, 683 = the maximum spectral luminous efficiency of radiation in lumens per watt, $1/4\pi$ (1 sr), and d = the distance from the light source (m). For example, assuming a zero value for reflectance, the illuminance or light intensity 2.4 m from a monochromatic orange light source (620 nm, where V_λ = 0.381 in Table 1.2) emitting 60W of power is:

60 x 0.381 x 683 x 0.0795 (1/4π) x 0.1736 (1/(2.4 x 2.4)) = 215 lx

For non-monochromatic light sources, illuminance can be calculated by summing the results of performing the above computation at 5-nm intervals between 400 and 780 nm (the range of wavelengths to which humans are sensitive). However, the photopic spectral sensitivity curve for poultry is different from that of humans, so the equivalent of illuminance for poultry, measured in units that have been variously termed *clux* or *gallilux*, could be calculated using the poultry spectral luminous efficiencies in Table 2 (interpolate for intermediate wavelengths).

Incandescence is the production of light by heating a material to a very high temperature. Solids used must have high melting points (>3000 K) and retain their shape at operating temperatures. Tungsten (melts at 3653 K, but operates at 2800 K in GLS vacuum or gas-filled lamps) and carbon (melts at 3825 K, but operates at 2100 K in filament lamps and 3800 K in carbon-arc lamps) are the most commonly used materials involving electrical energy. Candles and oil lamps use chemical energy when emitting light from glowing carbon particles within their flames.

Inverse-square law states that the illuminance (light intensity) at a point varies inversely with the square of the distance of the point from the light source. For example, if a light source radiates 32 lumens per steradian (lm/sr) of power, the illuminance at a distance of 1 m from the lamp will be 32 lux (or 32 lm/m²), because, by definition, a steradian subtends a spherical surface area equal to the square of its radius, so 32 lm of light will be spread over 1 square metre. At points 2 m from the light source, 32 lm of light will be spread over 4 square metres, producing an illuminance of 32/4 = 8 lx. So, at double the distance, the light intensity will be a quarter of the original value.

Irradiance (E_e) is the amount of radiant flux or power emitted per unit of surface area, and is measured in watts per square metre (W/m²). It is the radiometric equivalent of illuminance.

Light intensity is synonymous with illuminance and lighting level, and is expressed in lux.

Lighting level is synonymous with illuminance and light intensity, and is expressed in lux.

Lumen (lm) is the measurement of luminous flux emitted in a solid angle of 1 steradian (sr) by a light source with a luminous intensity of 1 candela (cd). It is the product of the maximum spectral luminous efficiency for photopic vision (683 lm/W), the radiant flux (W) and the luminous efficiency at the given wavelength ($V\lambda$). For example, a monochromatic blue light source emitting 100W at a wavelength of 450 nm will produce 683 x 100 x 0.038 (from Table 1.2) = 2595 lm.

Luminance (L_v) is the intensity of light emitted per unit area of a surface in a given direction, and is measured in candela per square metre (cd/m²). Surfaces, which can emit (lamp surface) or reflect (white wall) light will have a different luminance when viewed from different directions. It is the photometric equivalent of radiance.

Luminous efficacy is the ratio between the luminous flux (lm) emitted by a light source and its radiant flux (W). For example, if a 40W incandescent lamp produces 500 lm of luminous flux, its luminous efficacy is 500/40 = 12.5 lm/W. A 36W warm-white fluorescent lamp produces 2300 lm, five-times more light output than the incandescent lamp, and has a luminous efficacy of 63.9 lm/W.

Luminous flux (Φ_v) is the total quantity of light emitted by a light source per second, and is analogous to the volume of water going over a waterfall each second. It is synonymous with light power. Its photometric unit is the lumen (lm) and it is the product of the radiant flux or power (W) and the photopic luminous efficiency of the human eye (Vλ) at a given wavelength (λ) or range of wavelengths. At 555 nm wavelength (peak human photic sensitivity), 1 watt of radiant energy produces 683 lumens and, for example, the luminous power or flux of a 60W incandescent lamp is about 800 lm.

Luminous intensity (I_v) is the amount of light emitted in a given direction per unit of solid angle, and is measured in candela (cd) or lumens per steradian (lm/sr). When light is uniformly emitted in all directions, luminous intensity is equal to the luminous flux (lm) divided by 4π (steradians in a sphere). For example, if a 60W incandescent lamp emits 800 lm of light uniformly in all directions, the luminous intensity will be 800/4π = 64 cd.

Luminous power (Φ) is synonymous with luminous flux, and is measured in lumens (lm).

Lux (lx) is the measure of illuminance or light intensity, and, for photopic vision, is equal to 1 lm/m². It is equivalent to 0.09 foot candles or 0.08 cd/m².

Modulation is the number of light stimuli per second emitted by a fluorescent lamp, and is twice the frequency (Hz) of the alternating current supply. It is 100 Hz in Europe and 120 Hz in America.

Night is the period of darkness between dusk and dawn in which plasma melatonin release is elevated.

Phase reversal occurs when the shorter photoperiod in an intermittent lighting regimen is located in the latter half of the night and is interpreted as the beginning and not the end of the subjective day.

Phosphors are fluorescent powders used to coat the inner surface of fluorescent lamps. They convert the ultraviolet radiation produced by the mercury discharge of the lamp into longer wavelength visible radiation.

Photoinducible phase is the period within the internal biological cycle when the hypothalamus can be excited by light.

Photometric terms and units relate to the reception of radiation.

Photon is a packet or quantum of visible light.

Photon flux (N) is the total number of photons emitted per second, and has an influence upon spectral sensitivity. It is calculated from the radiant power (P), Planck's constant (h) and the photon velocity (c) using the formula:

$$N = P/h.c$$

where N = Photon flux per s, P = radiant power (W), h = 6.626 x 10^{-34} (J.s), and c = 2.998 x 10^8 (m/s).

Photoperiod (Greek, *photos* - light, *periodos* – circuit) is a period of illumination.

Photopic vision occurs when the eye is adapted to brighter light intensities and only involves photoreception by cones.

Plank's constant (h), 6.626 x 10^{-34}, is a figure by which the frequency of radiation (Hz) is multiplied to calculate the energy (Joules) of a quantum or photon of radiation.

Quantum is a packet of radiation that, if perceived as visible light, is called a photon.

Radiance (L_e) is the amount of radiant flux or power in a given direction and at a given point, and is measured in watts per square metre per steradian (W/m²/sr). It is the radiometric equivalent of luminance.

Radiant flux or Radiant power (Φ_e) is the amount of power emitted or received as radiation, and is measured in watts (W).

Radiometric terms and units relate to the emission of radiation.

Rods are one of two types of light-sensitive nerve endings in the retina of the eye, so called because of their shape. In the human eye, they outnumber cones, the other type of nerve ending, by more than ten to one and are highly light-sensitive. With the exception of the visual axis centre, an area called *fovea centralis (fovea)*, rods are evenly spread over the retina. Unlike cones, they do not have individual nerves; about a hundred are linked to a single nerve. Their numbers create a large light reception facility for maximising vision in dim light but, because many share the same nerve, the images produced are somewhat blurred. Colours cannot be distinguished with rods.

Scotoperiod (Greek, *scotos* - darkness, *periodos* - circuit) is a period of non-illumination and is synonymous with darkness.

Scotopic vision occurs when the eye is adapted to dim light intensities and only involves photoreception by rods.

Spectral luminous efficiency (V$_\lambda$) is the ratio of radiant flux at a given wavelength of light relative to that at maximum efficiency (Table 1.2), which for photopic vision in humans is 683 lm/W at a wavelength of 555 nm, and for scotopic vision is 1700 lm/W at a wavelength of 507 nm.

Speed of light or velocity of a photon (c, (Latin *celeritas* - speed) in a vacuum is the product of the wavelength (λ) of electromagnetic radiation and its frequency (v), and is constant at all wavelengths. It is approximately 3×10^8 m/s and can be calculated by the equation:

$$c = \lambda.v$$

where c = speed of light (m/s), λ = wavelength (m) and v = frequency (Hz).

Steradian (sr) is the solid angle whose projected spherical surface area (m²) is equal to the square of its radius (m). A sphere, therefore, contains 4π = 12.566 sr. Not an easy concept, but simplistically it is like 12.566 large wide ice-cream cones being put together to make a ball, with the angle of each cone being 1 sr, its length 1 m and the area of the cone opening 1 m².

Subjective day is the part of an intermittent lighting programme that is interpreted as day, and comprises various periods of light and darkness. Melatonin release is suppressed throughout this period.

Ultraviolet (Latin *ultra* – beyond, violet - the shortest wavelength of visible light) is the part of the electromagnetic radiation spectrum that lies between 200 and 400 nm and is invisible to the human eye. It is emitted by the sun as UV-A (long wave), UV-B (medium wave) and UV-C (short wave), but most UV-B and UV-C is absorbed by the ozone layer before it reaches the earth. It was discovered by the German physicist Johann Ritter in 1801.

UV-A, also called blacklight, is ultraviolet radiation between 320 and 400 nm. It is visible to some animals, including birds, and induces the production of vitamin D in the skin, but causes sunburn.

UV-B is ultraviolet radiation between 280 and 320 nm. It can cause molecular damage resulting in skin cancer.

UV-C is ultraviolet radiation between 200 and 280 nm. It has the highest energy of the three forms of ultraviolet and is the most dangerous, but can be used for sterilization.

Zeitgeber (German *zeit* – time, *geber* - giver) is an environmental agent or event that provides the cue for setting or resetting a biological clock, with light being the most important. The term was coined by the scientist Jürgen Aschoff.

• References •

1. Pritchard, D.C. (1995) The language of light, in: *Lighting*, pp. 1-14. Longman, Harlow.
2. Commission Internationale de l'Éclairage (1983) The basis of physical photometry. CIE, Vienna.
3. Prescott, N.B. & Wathes, C.M. (1999) Spectral sensitivity of the domestic fowl. *British Poultry Science* 40, 332-339.

2 PHYSIOLOGY AND MECHANISMS

This chapter describes the pathways of light transmission in poultry and the processes that take place after photoreception. Much of what we know about avian photobiology has been gained from species other than poultry, especially the quail, but the principles apply equally to most species. Avian and human visions are contrasted, and the anatomical differences that affect our respective perceptions of light from different sources are discussed. The abbreviations d, h and min are used for days, hours and minutes respectively and acronyms such as 14L:10D (for example) are used to represent a cycle of 14 h light and 10 h darkness.

• Avian vision and comparative luminous efficiencies •

Birds have relatively large eyes in comparison to humans, with the combined weight exceeding that of the brain. Their flattened shape gives them the functional ability of much larger eyes whilst minimising their weight and volume. The location of the eyes on the side of the head provides the bird with small central binocular, but wide monocular, fields of view. Although poultry are generally neither long nor short-sighted, they are increasingly short-sighted in their lower field of vision, and this may allow the bird to keep near and distant objects in focus simultaneously – a useful facility that allows the bird to remain alert while foraging for food.

There are two types of photoreceptor cell in the retina of the eye: rods, which are more numerous, highly sensitive and allow vision in poor light (scotopic conditions) and cones, which are responsible for normal daytime (photopic conditions) vision. The image produced by rods is poorly defined because a large number are linked to one nerve fibre. However, a summation of stimuli gives rods their high sensitivity, and this is maximal at 507 nm (blue-green light). Rods, which are only operational below 4 cd/m², are unable to distinguish colours. In contrast, the less numerous cones, each with its own nerve fibre, are responsive at much brighter levels of illuminance (starting from 4 cd/m² and maximal from about 44 cd/m²), produce high definition images, and allow the perception of colour. In the human eye, cones are responsive to electromagnetic radiation between approximately 400 and 730 nm, with a maximum sensitivity at 555 nm. There are three types of cone and each is capped with a particular oil droplet that allows specific wavelengths of light to penetrate through to the connecting nerve fibre. The peak sensitivities of the three types of cone allow us to perceive the primary colours, violet/blue (450 nm), green (550 nm) and red (700 nm) When all are stimulated simultaneously the brain registers the incoming light as white. Avian eyes have an additional type of cone in the retina with a peak sensitivity at about 415 nm (1, 2), and these cones allow the detection of radiation below 400 nm (3). Additionally, the lens and humours of the bird's eye are optically clear between 320 and 400 nm (1, 2), and this means that poultry can 'see' in the UV-A part of the ultraviolet range.

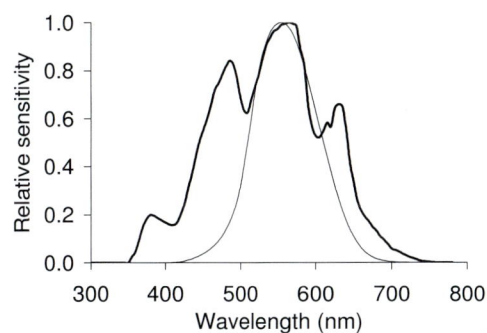

Figure 2.1 *Relative spectral sensitivities of domestic fowl (thick line) normalised to a sensitivity of 1.0 at 565 nm (3) and humans (thin line) normalised to a sensitivity of 1.0 at 555 nm.*

As a consequence, poultry probably perceive colour differently from humans. Although birds have a fourth type of cone, peak avian sensitivity occurs in a part of the spectrum similar to that in humans (545 to 575 nm) (3). However, the spectral sensitivity of birds between 400 and 480 nm and between 580 and 700 nm is greater than that of humans (Figure 2.1, Table 1.2, page 2). This must result in poultry perceiving light from some lamps as brighter than humans, but the degree of extra brightness will vary with the type of lamp. It therefore follows that the lux (a photometric unit calculated from the spectral power output of a light source and the sensitivity of the human eye) is not strictly appropriate for describing levels of illuminance in poultry houses. Theoretical perceptions by humans and poultry of illuminance from various types of lamp are given in Table 2.1. The illuminance for poultry, variably termed *clux* or *gallilux*, at a given distance from a light source was obtained by integrating the illuminance (*I*) for 5 nm segments of the spectrum and is calculated as follows:

$$I = (w.s.683)/(12.566.d^2)$$

where w is the power output of the lamp (W) in 5 nm segments, s is the relative sensitivity of domestic fowl, 683 is the maximum luminous efficacy for human photopic vision (lumens/W), 12.566 (4π) is the number of steradians in a sphere and d the distance from the lamp (m). The maximum spectral luminous efficacy of radiation for photopic vision for humans was used for both calculations because no avian data are available. However, if the figure for poultry is not 683 lumens/W, then the absolute values in Table 2.1 will be incorrect, although the relative values given for *gallilux* are still valid. If the avian efficacy is markedly less than 683 lumens/W, the suggestion that poultry perceive light more brightly than humans would be incorrect. Indeed, if birds perceive the intensity of sunlight to be similar to humans (i.e. the area under the spectral sensitivity curves for birds and humans are the same), the maximum spectral luminous efficacy for poultry would be only 400 lumens/W.

Table 2.1 *Perceived illuminance by humans and domestic fowl at a distance of 1.5 m from light sources, with values calculated using spectral sensitivities, lamp power, a maximum spectral luminous efficacy of radiation for photopic vision of 683 lumen/W and a reflectance value of 0.2.*

Light source	Irradiance (W/m²)	Intensity perception Human (lux)	Intensity perception Fowl (gallilux)	Ratio of fowl to human illuminance
15W incandescent lamp	0.03	5.6	8.1	1.45
Warm white fluorescent tube	0.28	120.8	147.2	1.22
Cool white fluorescent tube	0.30	120.8	159.1	1.32
70W Sodium high pressure lamp	0.52	254.4	277.3	1.09
36W Blacklight/blue lamp	0.28	0.7	31.1	41.86
36W Blue fluorescent tube	0.42	37.8	196.8	5.20
36W Red fluorescent tube	0.03	2.2	6.7	3.05
Summer sunlight at noon and cloudless sky in UK (4)	487	100,000	163,560	1.64

• Photoreception and responses to illumination •

Retinal reception

There are two routes by which light reaches the brain to modify the behavioural and sexual activity of poultry. The first and most obvious is through the eye, and it is this illumination that allows the bird to see. Light is absorbed by photopigments in the retina (rhodopsin in rods and iodopsin in cones) to form inverted images that are converted into complex electrical signals for transmission to the brain *via* the optic nerve. However, the processes by which photons of light energy are converted into neural signals are still not fully understood. Many behavioural and biological responses are dependent upon retinal photoreception. In addition, light entering the eye stimulates the synthesis, release and metabolism of dopamine, which, in turn, suppresses the production of serotonin-N-acetyltransferase, the main enzyme involved in the regulation of melatonin production in the retina during the period of 'night-time' darkness. This dopaminergic neurotransmission also suppresses melatonin biosynthesis in the pineal gland, especially at low light intensities (5, 6). Melatonin synthesis in the pineal can also be suppressed by signals generated by retinally received UV-A radiation, a process that, in contrast to white light reception, does not involve dopamine but is activated by stimulation of retinal N-methyl-D-aspartate (NMDA) glutamate receptors (65).

Furthermore, dopamine prevents the release of enkephalins, neurotensin and somatostatin, agents that also appear to be involved in the control of melatonin production in the eye. At low light intensities ($\leq$ 4 lux), when it is too dim for light to penetrate bone and cranial tissues, retinally released dopamine also suppresses serotonin and melatonin synthesis in the pineal gland (6). However, this process only operates within a very narrow band of light intensity because, below 0.1 lux, dopamine ceases to be produced, and serotonin secretion and melatonin synthesis are reactivated.

Pineal reception

In birds, the pineal gland is located on the top of the brain in the triangular space between the cerebral hemispheres and the cerebellum (Figures 2.2 and 2.3, 7), and, in contrast to humans, has both a circadian clock and functional photoreceptors (8). When illuminance is brighter than 4 lux, light is able to pass directly through the skull and cranial tissues to the pineal gland, where it suppresses the production and release of serotonin and melatonin (10).

Hypothalamic reception and photosexual response

The hypothalamus, also referred to as the extra-retinal or deep encephalic photoreceptor, is a collection of nuclei situated in the preoptic part of the forebrain (Figure 2.3). Light that reaches this area directly through the skull and tissues, or which induces a neural signal from the retina to the suprachiasmatic nucleus (SCN), controls the secretion of gonadotrophin releasing hormone (GnRH). This hormone is transported *via* the hypophyseal portal blood vessels to the anterior pituitary gland where it stimulates the release of luteinizing hormone (LH) and follicle stimulating hormone (FSH). The amount and timing of gonadotrophin release influences the

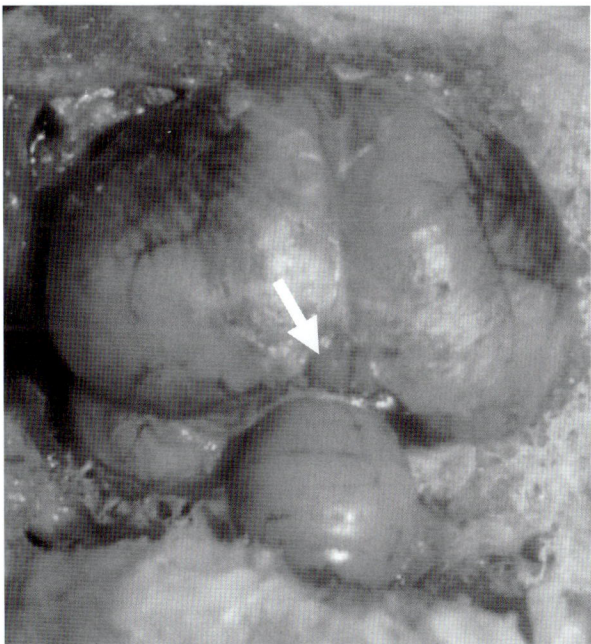

Figure 2.2 *Dorsal view of the brain showing the pineal gland (indicated by the white arrow) located between the cerebral hemispheres (upper) and the cerebellum (lower) on the top of the brain.*

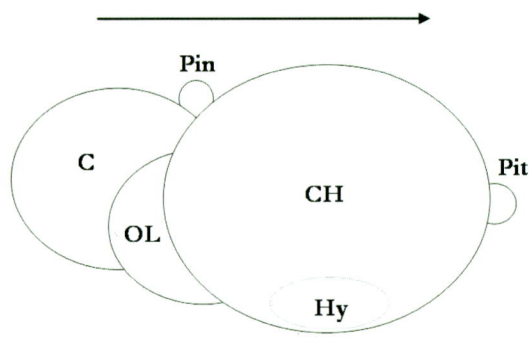

Figure 2.3 *Schematic diagram of the avian brain showing the relative locations of the cerebellum (C), optic lobe (OL), cerebral hemisphere (CH), pineal gland (Pin), pituitary gland (Pit) and the hypothalamus (Hy), with the arrow pointing forward.*

rate of sexual maturation. Recently, a further hypothalamic hormone, gonadotrophin inhibitory hormone (GnIH), has been identified, and it is suggested that this may play an important role in the control of LH and FSH releases that are involved in sexual maturation and in the termination of egg-laying in seasonal breeding birds (11). Additionally, expression of the gene that encodes the enzyme responsible for catalysing the conversion of thyroxine (T_4) into triiodothyronine (T_3) within the hypothalamus, type 2 iodothyronine deiodinase (Dio2), has been shown to be induced by photostimulation (12). Evidence to support a role for Dio2 in the initiation of sexual maturation is the observation that the administration of the Dio2 inhibitor, iopanoic acid, prevents gonadal development. Thyroid hormones play an important role in the processes that dissipate juvenile and trigger adult photorefractoriness in turkeys (13). The suppression of T_4 production, either pharmacologically (14) or by thyroidectomy (15, 16), also has a major influence on the inhibition or extension of reproductive function in male and female turkeys.

Fos, one of the genes that couples short-term signals received at a cell surface to long-term changes in cellular response, may also be involved in inducing GnRH release (17). Fos expression has been demonstrated to occur in the median eminence and infundibular nucleus of the mediobasal hypothalamus within 18 h of a transfer to a stimulatory photoperiod, and before any rise in LH. Fos-like immuno-reactivity was observed in neuroglia, the connective tissue that surrounds and supports nerve cells, within the median eminence and in the neural cells themselves within the infundibular nucleus.

Although the eye is not essential for light stimulation of the hypothalamus in poultry (18), it may, as for the pineal gland, still be the primary site of light reception at low intensities and, as a consequence, be relevant in controlled environment poultry houses, where low illuminance is frequently used to control bird behaviour.

Table 2.2 *Estimates of the mean relative transmission of light of a given colour passing through avian skull and brain tissues to reach the basal hypothalamus.*

Human colour sensation and wavelength (nm)	Transmission relative to red (650 nm)			
	Quail (21)	Sparrow (22)	Pigeon (22)	Ducks (20)
Violet (400-435)	0.013	0.018	0.275	
Blue (435-500)	0.023	0.055	0.098	0.027
Green (500-565)	0.020	0.067	0.113	0.096
Yellow (565-600)	0.090	0.135	0.158	0.244
Orange (600-630)		0.271	0.280	0.410
Red (650)	1.000	1.000	1.000	1.000
Red (700)		2.927	11.681	

• Penetrability to, and sensitivity and intensity thresholds at, the hypothalamus of different wavelengths of light •

In addition to the difference between the efficiency with which light activates the hypothalamus *via* the retina and optic nerve and the direct route through the skull and cranial tissues, there is variation in the efficiency with which different wavelengths of light penetrate directly to the hypothalamus. When Mallard drakes were exposed to monochromatic light, with peaks ranging from 436 to 836 nm, testicular development only occurred under 617 nm (orange), 664, 708 or 740 nm (red) wavelengths of light (Figure 2.4), suggesting that the threshold wavelength for light to successfully penetrate to the hypothalamus and effect a photosexual response was somewhere between 577 and 617 nm (19). However, it was subsequently demonstrated that although light absorption increases as wavelength decreases, some light still reaches the hypothalamus at wavelengths as short as 400 nm (Table 2.2). The penetration of longer wavelength red light (>650 nm) to the hypothalamus has been variously reported to be 36 times for ducks (20), 40 times for quail (21) and between 100 and 1000 times for sparrows. Although the original research (19) suggested that testicular growth in Mallard drakes was only possible under wavelengths of light longer than 617 nm, it was subsequently shown that the lack of development under the shorter wavelengths was due to differences in the degree to which light was able to penetrate the intracranial tissues (23). Testicular development in drakes which had had their optic nerves severed was similar when red,

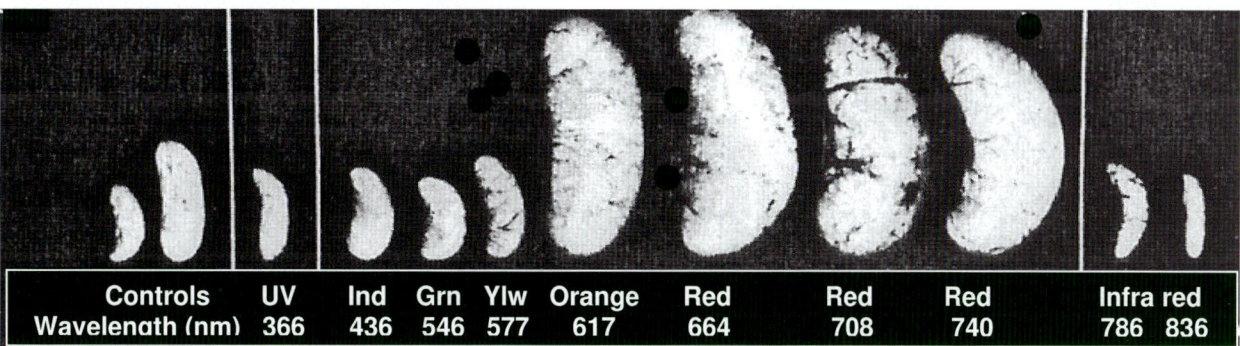

Figure 2.4 *Testicular growth in Mallard drakes following conventional exposure to various wavelengths (nm) of illumination: ultraviolet (UV), indigo (Ind), green (Grn), yellow (Ylw), orange, red and infra red (19).*

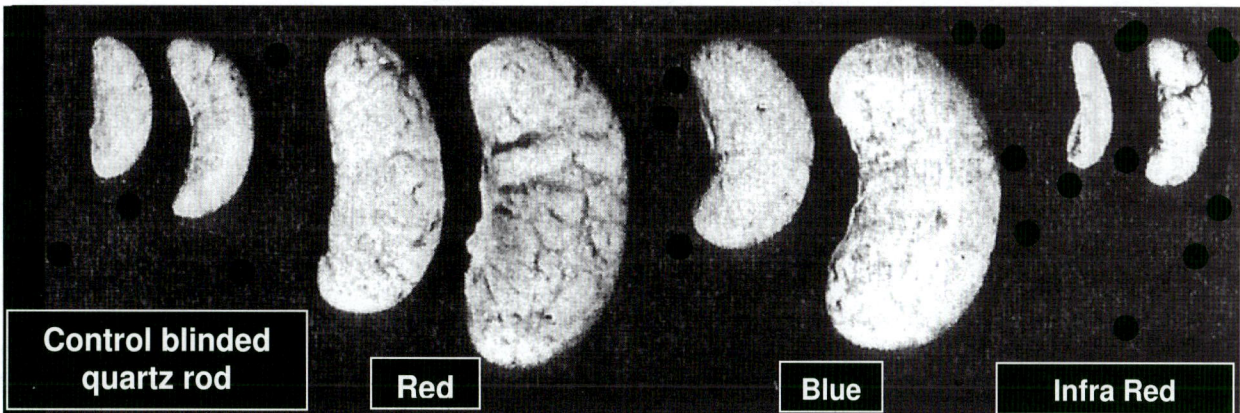

Figure 2.5 *Testicular growth in Mallard drakes following exposure to red light, blue light or infrared radiation transmitted directly to the hypothalamus through a quartz rod (23).*

Table 2.3 *Transmission efficiencies for the passage of light from a quartz-halogen lamp through the skull and brain tissues to the hypothalamus, sensitivity of hypothalamic photoreceptors and estimates of the anticipated photosexual response to various wavelengths of light for quail as measured by a rise in plasma LH concentration (21)*

Wavelength (nm)	Transmission efficiency T/T_{max} [†]	Relative to 470 nm	Sensitivity [‡] Photons^{-1}	Relative to 650 nm	Response Relative transmission x relative sensitivity	Relative to 470 nm
470	6.7×10^{-6}	1.0	3.02×10^{-11}	11.8	11.8	1.0
500	7.1×10^{-6}	1.1	5.83×10^{-11}	22.8	24.4	2.1
590	2.4×10^{-5}	3.7	1.11×10^{-11}	4.3	15.9	1.3
650	2.7×10^{-4}	40.5	2.56×10^{-12}	1.0	40.5	3.4

[†] T_{max} is the maximum transmission of light (photons/m².s) that occurs when no tissue is placed between a light source and the detector, and T is the actual transmission detected at the hypothalamus.
[‡] Sensitivity is the reciprocal of the minimum light intensity required at the hypothalamus to induce a change in plasma LH concentration.

yellow, indigo or blue light was directed immediately on to the hypothalamus via a quartz rod (Figure 2.5). In contrast, whereas blue-green light (500 nm) that had been directed to the hypothalamus *via* optic fibres at an intensity of 1.71×10^{10} photons/m².s was sufficient to elicit a rise in plasma LH concentration in castrated quail, twice as much violet (470 nm), five times as much yellow (590 nm) and 23 times as much red light (650 nm) was needed to induce the same response (21). Although this might suggest a difference in sensitivity between ducks and quail, the varying results could have been due to the different techniques used to direct light to the hypothalamus.

Notwithstanding that avian hypothalamic photoreceptors are most sensitive to blue-green (500 nm) light (at least in quail) when it is transmitted directly to the hypothalamus, conventional illumination by red light will still induce a bigger photosexual response than any other colour of light at the same photon flux. This is because the more efficient penetration to the hypothalamus by longer wavelengths of light will more than compensate for their lower sensitivity at the hypothalamic receptor (Table 2.3).

In addition to the influence that it has on the photosexual response, experiments have shown that wavelength of light can modify a bird's ability to do simple tasks when given at different intensities (26). Domestic pullets which had been trained to take a feed reward from either a brightly or a dimly illuminated chamber were unable to discriminate between dim red (635 nm) and bright violet (415 nm) light when the intensity of red light was increased to 3.1×10^{22} photons/m².s and the violet was reduced to 11.2×10^{22} photons/m².s (3.6 times red). When the reversed starting conditions (bright red and dim violet) were tested, colour discrimination failed with red light at 2.5×10^{22} photons/m².s and violet at 7.8×10^{22} photons/m².s (3.1 times red). These experiments show that, as judged by their ability to do a simple task, chickens see red light as being equal in illumination to about three times the quantum flux of violet light.

The threshold of white light sensitivity for testicular enlargement in immature intact Mallard drakes has been reported to be between 1 and 5 lux (27). This sensitivity decreased by about five-fold when the optic nerve was severed, thus limiting photoreception to the external stimulation of the hypothalamus. Notwithstanding that the lux is not an appropriate unit of illuminance for birds, these data indicate the relative sensitivities of the retina and hypothalamus. The threshold intensity at the skull surface necessary to maximise the photosexual response (largest change in plasma LH concentration) in immature quail is about 6×10^{16} photons/m².s, although some individuals have been observed to respond at an intensity of 3.9×10^{15} photons/m².s (21).

• Photoinducible phase •

Birds are only photosexually responsive to light for a limited period of the light-dark cycle, and this is termed the photoinducible phase (PIP) (28). The first evidence of a circadian rhythm in avian photosensitivity came from an experiment in which immature male house finches were given 6 h of illumination in light-dark cycles of 12, 24, 36, 48, 60 or 72 h (29). After 22 cycles of treatment, testicular growth had only occurred in birds given 6 h of light within a 12, 36 or 60-h cycle, and then it was a graded response, with those subjected to more frequent illumination having heavier testes. Where the 6-h photoperiod was followed by 18, 42 or 66 h of darkness, there was no enlargement of the testes (Figure 2.6). Similar variations in the rate of sexual maturation were observed in male and female Japanese quail that had been exposed to the same lighting schedules as those used for the finches (30). The results indicated a circadian rhythm, with the PIP lying somewhere in the subjective night.

Testicular growth rates in quail maintained on a 6L:18D cycle and given 15-min pulses of light at various times in the dark period suggested that the PIP lasts for about 4 h and lies between 12 and 16 h after the start of the main photoperiod (30) (Figure 2.7). These timings were confirmed in an investigation in which castrated quail (castration eliminated gonadal hormone feedback) kept on 8L:16D cycles were given a single 4-h light pulse at various times during the dark period (31). Plasma LH concentration only increased when a pulse was given between 10 and 14 h, 11 and 15 h or 12 and 16 h after dawn, with the highest response induced by a 12-16 h pulse.

The location of the PIP in quail is determined by the timings of dawn and dusk and by the size of the main photoperiod, and has been estimated to be delayed by 20 min for each 1-h extension of the photoperiod (32). The effect of photoperiod on the position of the PIP has also been demonstrated in domestic pullets (33). When pullets were give 15-min light pulses in the dark period of a 7.75L:16.25D cycle, pulses given between 14 and 16 h after dawn were the most effective for elevating plasma LH, but, when pullets were given 4-h pulses in the night of a 4L:20D cycle, maximum sensitivity occurred about 11 h after the start of the photoperiod. However, the peak response occurred at the same time relative to the end of the photoperiod, suggesting that dusk, and not dawn, is the main entraining cue for PIP in domestic fowl.

Differences in the rate of testicular growth for quail given different length light pulses at 5-10 lux indicate that the size of the photosexual response might be proportional to the amount of PIP illuminated (34). However, other studies with quail have demonstrated that the intensity of the light pulse is also important and may question the need for complete illumination for maximum response, because a 1-h pulse of light at an intensity of 850 lux and starting 7 h into the dark period of a 6L:18D cycle gives

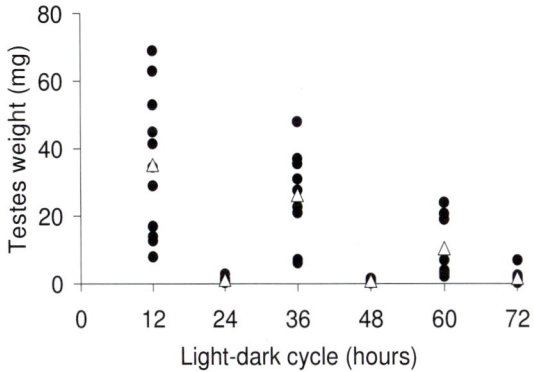

Figure 2.6 *Individual (●) and mean (△) testes weights for house finches exposed to 6-h photoperiods within light-dark cycles of 12, 24, 36, 48, 60 or 72 h (29).*

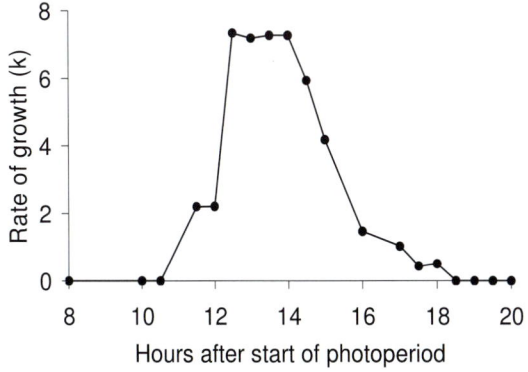

Figure 2.7 *Rate of testicular growth in quail given a 15-min light pulse at various times in the night of a 6L:18D cycle (30).*

significantly greater testicular growth than the same size pulse at 250 lux (35).

There is also a difference between the amount of light required within the PIP to initiate gonadal growth and that needed to sustain gonad size in mature birds (35). Whereas a 15-min light pulse located 7.5 h into the night of a 6L:18D cycle is sufficient to maintain testicular mass in quail, a 2-h pulse is required to maximise testicular growth.

• Phase reversal •

Phase reversal can occur when light pulses are used in night interruption experiments and in commercial intermittent lighting programmes when the period of darkness that follows the main photoperiod is discernibly longer than the period of darkness that comes after the shorter photoperiod. Normally, when the light pulse is given in the first half of the night, the bird will interpret the start of the main photoperiod as dawn and the end of the pulse as dusk, and the period in between as day. For example, 7L:3D:4L:10D will be seen as a 14-h subjective day (7+3+4) and a 10-h night, with the PIP illuminated by the 4-h pulse. However, if the pulse is given in the latter half of the night, such as 7L:8D:4L:5D, then the beginning of the 4-h pulse will be seen as dawn and the end of the 7-h photoperiod as dusk, thus forming a 16-h subjective day (4+5+7) and an 8-h night. This regimen should really be described as 4L:5D:7L:8D because the shorter photoperiod is now the beginning of the subjective day. Any circadian rhythm that is controlled by dusk, such as locomotor activity and the ovulatory cycle, will be phase shifted because its zeitgeber has moved.

The point where phase reversal takes place depends on the length of the main photoperiod - occurring later with longer photoperiods (36) - and on the duration and intensity of the light pulse (32).

• Photoperiodic time measurement •

Birds are equipped with a very sophisticated mechanism for measuring daylength and changes in daylength, and this must be particularly so in annually migrating species. Many theories have been put forward to explain this mechanism, but a universally accepted hypothesis has yet to be offered. The earliest suggestion, which involved an 'hour glass' or 'interval timer' principle, is called 'Bünning's hypothesis' (37). This envisaged an undefined photochemical process occurring during either the light or dark period that was reversed in the opposite phase of the daily cycle. When enough product of the reaction had accumulated by exposure to a long enough light period a threshold was thought to have been exceeded and a physiological process initiated. A development from this simple theory was the 'external coincidence' model in which an endogenous rhythm of cellular activity was comprised of two half cycles per day, one light requiring and the other darkness requiring, and a sinusoidal circadian rhythm of photosensitivity (38). The positive section of the curve denoted the photosensitive period, which occurred during the subjective night, and the negative section corresponded to the photo-insensitive subjective day. If illumination occurred during the 12 h photosensitive part of the cycle a sexual response would be induced whose magnitude was a function of the duration of the coincidence. Entrainment was effected by the complete light cycle, not just dawn or dusk, and therefore the phase angle of the median point of the positive section of the curve (sine maximum) moved with different light-dark ratios.

In contrast, the 'internal coincidence' model envisaged light having only one function, namely that of entrainment of circadian rhythms (39). Time was measured by the photoperiod entraining two circadian oscillators, each of which had its own entrainment properties. Changes in the light-dark cycle altered the phase angle between the two oscillators to an extent that when a critical photoperiod was experienced a mutual phase angle existed and photoinduction occurred. It was postulated that one of these oscillators was entrained by dawn and the other by dusk.

The unknown photochemical postulated in Bünning's hypothesis is melatonin, the hormone synthesised in the retina and pineal gland during night-time darkness. In the house sparrow, the nocturnal melatonin signal is long but of low amplitude after short photoperiods, and short but high in amplitude following long photoperiods (40). This information is encoded and thought to be biologically stored in the hypothalamus to control circannual activity. Transfers from short to long-days result in a four-fold increase in the number of neurones in the mediobasal hypothalamus of quail in which there is Fos-like immunoreactivity (41). This activity, which begins early in the second long-day, continues for at least three more days and is associated with an increase in LH release, suggests a role for the neurones in this part of the avian hypothalamus in the transmission of photic information to the cells that secrete GnRH.

In studies with quail, it has been shown that several circadian clock genes in the suprachiasmatic nucleus (SCN) of the hypothalamus, the likely site of the circadian clock, and in the pineal gland express differently under 8L:16D and 16L:8D cycles (42). Phase angles for peak expression of *Cry1, Cry2, E4bp4,* and *Per2* genes are significantly different in the SCN, and for *Bmal1, Per3, Clock, Cry1,* and *E4bp4* in the pineal gland. However, the differences between short and long-days are not the same as those for birds held on 4-h photoperiods and given 30-min light pulses at 3 or 17 h (PIP not illuminated), or at 10 h (PIP illuminated) into the 20-h dark period. There are two peaks of expression for *Per2* and *Cry1* genes in the SCN, and significantly higher mRNA levels for *Clock* and *E4bp4* genes in the pineal gland for birds given the stimulatory light pulse than for those on non-stimulatory schedules. The birds on the stimulatory schedule also have significantly heavier testes weights after 10 cycles of night interruption lighting. The difference between the responses to short and long-days and those to stimulatory and non-stimulatory interrupted lighting regimens show that photoperiodic time measurement in quail, or maybe birds in general, is complex. A gene, whose expression is only induced when a stimulatory light pulse is given to the mediobasal hypothalamus, has recently been identified, and this might be responsible for photoperiodic time measurement (42).

• Carry-over effect •

More than 50 years ago, it was postulated that the avian photoinduced reproductive system responded almost immediately a photoperiod began and that there was a carry-over effect of the illumination after the photoperiod had ended (43). Subsequently the validity of this 'pendulum' or 'fly-wheel' effect was confirmed in white-crowned sparrows, house sparrows, quail, and domestic fowl.

Observations from deafferentation experiments with quail indicated that the basis of the effect lies in the hypothalamus and not in the pituitary gland (44). Figure 2.8 shows the dependence of the carry-over effect upon previous lighting history and on the degree of maturity of the birds (44). The modifying effect of amount of illumination was shown in a trial

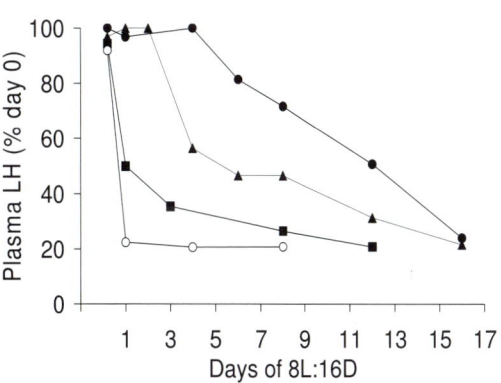

Figure 2.8 *Plasma LH concentration normalized to 100% for 0d, for quail transferred from stimulatory to 8-h photoperiods: mature birds* (●), *after 4 cycles of 20L:4D* (▲), *after 2 weeks of intermittent lighting* (■) *and following hypothalamic deafferentation* (○) *(44).*

Table 2.4 *Relative daily feed intake for laying hens given various mixtures of 8L:4D:2L:10D and 2L:22D lighting compared with permanent 8L:4D:2L:10D controls (48).*

Period (d)	Regimen[†]	Feed intake (g)	Regimen[†]	Feed intake (g)
1-14	2(1S:6L)	-10.0	2(2S:5L)	-17.4
15-32	3(1S:5L)	-9.1	3(2S:4L)	-18.5
33-47	3(1S:4L)	-1.1	3(2S:3L)	-3.4
48-63	4(1S:3L)	-11.7	4(2S:2L)	-9.5
64-75	4(1S:2L)	-5.9	4(2S:1L)	-19.0
76-87	6(1S:1L)	-7.7	3(3S:1L)	-15.8
88-97	10S	-23.4	2(4S:1L)	-11.1

[†] S = 2L:22D, L = 8L:4D:2L:10D

with white crowned sparrows. Six 50-min light pulses given equally throughout the 24 h (5 h illumination) induced a testicular growth rate equal to a 13L:11D conventional regimen, whilst 12 equally spaced pulses (10 h illumination) were as stimulatory as a 17L:7D schedule (44).

The carry-over effect has also been demonstrated by mixing short and long-days (44, 45). For example, ovarian and oviducal weights for quail normally exposed to 8-h photoperiods but given a 16-h photoperiod every second or fourth day were not significantly different from those of birds transferred permanently to 16-h photoperiods. Combined organ weights for birds given a 16-h photoperiod every sixth day were intermediate between constant 8-h controls and birds transferred to 16 h.

The carry-over phenomenon was also observed in sparrows transferred from stimulatory daylengths to continuous darkness; testicular size was unaffected after 4 weeks of treatment (46).

The persistency of a photoperiodic influence has been reported in both growing pullets and laying hens. The provision of a novel 'saw-tooth' lighting programme from 2 d in which the pullets were given repeated daily increases of 30 min from 8 to 14 h photoperiods followed by an abrupt 6-h decrease back to 8 h resulted in mean age at first egg being similar to constant 14-h controls, 1 day earlier than constant 8-h controls but 7 d later than birds maintained on 11-h photoperiods (47). This showed that the effects of the 14-h photoperiod in the 'saw-tooth' programme, though only given once every 14 d, continued for the next 13 d to be interpreted as a constant 14-h programme.

In laying hens, neither egg production nor egg weight was affected when the main 8-h photoperiod was removed from an 8L:4D:2L:10D intermittent schedule at progressively more frequent intervals between 49 and 63 weeks of age to reach either a daily 2-h photoperiod only or two cycles of 4 d of 2 h followed by one day of 8L:4D:2L:10D in the final 10 d (48). However, feed intake decreased and feed conversion efficiency improved as more days with only 2 h illumination were inserted (Table 2.4).

• Photorefractoriness •

Photorefractoriness is simply the inability to respond sexually to an otherwise stimulatory daylength. The condition, which is in some way facilitated by T_4 hormone (13, 49), is dissipated in birds not on restricted feeding by exposure to about two months of non-stimulatory photoperiods. These are provided in nature by short winter daylengths so that sexual maturation can occur early in the following spring. Alternatively, it can be terminated by markedly reducing illuminance when birds are held on long days (50, 51) or by pharmacologically inducing hypothyroidism (14). There is also a possible involvement of prolactin, although this seems to be species-specific in its association. Whereas plasma prolactin increases during the onset of adult photorefractoriness in starlings (52), elevated prolactin is coupled with the persistence of photosensitivity in turkeys (13).

Seasonal breeding birds are hatched photorefractory, and this juvenile form of the condition prevents them from becoming sexually mature in their first year, even though they may be somatically mature. The adult form terminates reproduction in sexually mature birds so that offspring are not being reared when environmental conditions are unfavourable. Truly seasonal breeding birds, like pheasant and partridge, exhibit an absolute form of photo-refractoriness and have a short breeding season of 2 to 3 months. Once photorefractory, they remain infertile until the following year and some individuals, if held on long-days from hatch, might never become sexually mature. In one experiment in which red-legged partridge were held on 16-h photoperiods, the first bird did not lay until it was 68 weeks of age and only 38% of birds had reached sexual maturity by 215 weeks (53).

Broiler breeders exposed to stimulatory photoperiods during the rearing period mature 3 to 4 weeks later (54) and produce 15-25 fewer eggs (55, 56) than those given non-stimulatory daylengths. Additionally, if broiler breeders or turkeys are transferred to stimulatory photo-periods before they have become photo-sensitive, they will not only fail to be photo-stimulated but will respond to the long day as if they had always been on it and so have their maturity markedly delayed (56). When a flock is photostimulated after the first but before the last individual bird has fully dissipated photorefractoriness, there will be a bimodal distribution and a wide range of ages at first egg resulting in a low peak in egg production and variable egg weight.

Unpublished data from three trials conducted at the University of KwaZulu-Natal, in which broiler breeder pullets were grown to reach 2.1 kg at about 20 weeks and transferred from 8 to 16-h photoperiods at ages ranging from 10 to 25 weeks, suggest that juvenile photorefractoriness is not fully dissipated from a control-fed flock until at least 18 weeks of age (Figure 2.9). This is about double the figure suggested by the earlier research conducted with quail and sparrows (57). In other unpublished work conducted at the University of KwaZulu-Natal, broiler breeders that had been grown to reach 2.1 kg at 20 weeks did not fully respond to photo-

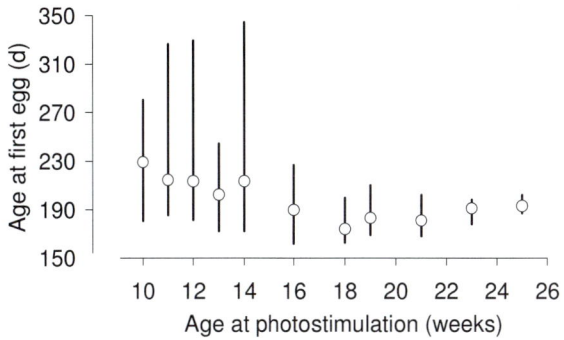

Figure 2.9 *Range (bars) and mean (○) age at first egg for broiler breeder pullets transferred from 8 to 16 h at various ages between 10 and 25 weeks (unpublished data from the University of KwaZulu-Natal).*

stimulation until 18 weeks, but when feed-control was relaxed to achieve 2.1 kg by 15 weeks, the flock responded from 14 weeks. Quails and sparrows in the earlier trials would have been fed *ad libitum*; and so it is likely that the slower rate at which photorefractoriness is dissipated in broiler breeders is due to a delaying effect of controlled feeding (56).

When birds are given constant daylengths, photorefractoriness was thought to be dissipated naturally at a rate that was inversely related to the photoperiod, i.e. the longer the daylength the longer it took for the bird to be photoresponsive (58). However, in broiler breeders it is far from being a simple linear effect, with the earliest sexual maturity occurring for birds held on 10-h photoperiods, the latest, which presumably indicates the slowest dissipation of photorefractoriness, between 13 and 14 h, followed by a progressively quicker rate of dissipation for longer photoperiods (54). This suggests that the rate of dissipation is not dictated by the duration of illumination, but by the stimulatory nature of the photoperiod.

The rate at which adult photorefractoriness develops is correlated with the size of the photoperiod given during the laying period (59). However, season and age also seem to be involved in determining the onset of adult photorefractoriness. It occurred earlier in turkeys photostimulated in the spring (17 weeks) than in the winter (25 weeks), and less than 90% of first-year turkey hens were observed to become photorefractory whilst all ceased laying in the second cycle (60).

Photorefractoriness poses an additional problem in turkeys and, perhaps, broiler breeders in that a flock is likely to contain birds that exhibit the adult form to varying degrees (61); some will still be photosensitive at depletion, some will have become photorefractory while others may have gone out of lay and spontaneously resumed egg production despite being maintained on 16-h photoperiods (Figure 2.10, 62). In contrast, photorefractoriness is minimal in modern egg-type hybrids. Differential rates of response by broiler breeders and egg-type pullets to photoperiods longer than 10 h indicate that juvenile photorefractoriness only affects sexual maturation in broiler breeders (Figure 2.11). Whereas sexual maturity is significantly delayed in broiler breeders when they are maintained on stimulatory ($\leq$ 11 h) photoperiods (54), sexual development in modern egg-type hybrids is hardly affected at all (63). There also is no evidence to indicate that modern egg-type hybrids exhibit adult photorefractoriness (64). Figure 2.12 shows that the profile of egg production for birds transferred from 8 to 11 h was similar to that for hens given a step-up programme from 8 h to reach a 15-h photoperiod at 27 weeks of age.

The almost total elimination of photo-refractoriness from egg-type stock is likely to be a consequence of the more intense selection for egg numbers within a layer breeding programme compared with that for meat-type stock, where the emphasis is on growth rate, conformation and carcass composition.

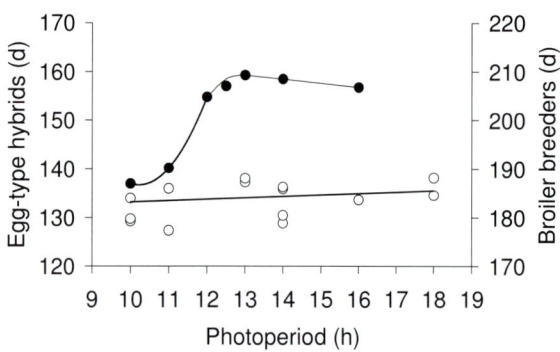

Figure 2.11 *Mean age at sexual maturity for modern egg type (○) and broiler breeder (●) pullets maintained from 2 d on photoperiods ranging from 10 to 18 h (54, 63).*

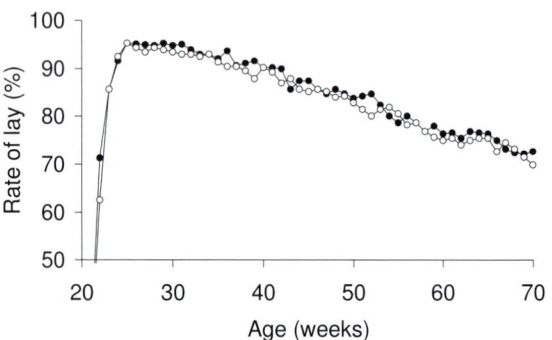

Figure 2.12 *Rate of lay for hens transferred from 8 to 11 h at 20 weeks (○) or given 1-h increments from 18 weeks and then 30-min increments to reach 15 h at 27 weeks (●) (64).*

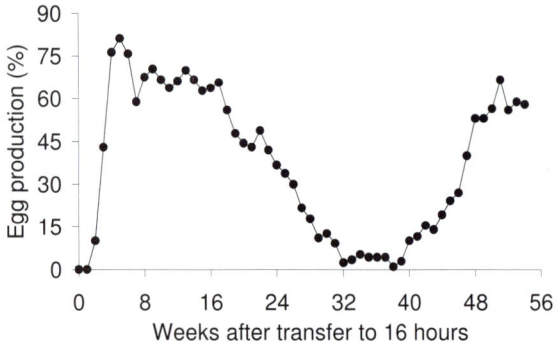

Figure 2.10 *Egg production for turkeys recycled with a period of 8-hour days and then transferred permanently to 16-hour days (62).*

Chapter 2 Physiology & mechanisms

• References •

1. Govardovskii, V.I. & Zueva, L.V. (1977) Visual pigments of chicken and pigeon. *Vision Research* 17, 537-543.
2. Hart, N.S., Partridge, J.C. and Cuthill, I.C. (1999) Visual pigments, cone oil droplets, ocular media and predicted spectral sensitivity in the domestic turkey (Meleagris gallopavo). *Vision Research* 39(20), 321-3328.
3. Prescott, N.B. & Wathes, C.M. (1999) Spectral sensitivity of the domestic fowl. *British Poultry Science* 40, 332-339.
4. Monteith, J.L. (1973) The radiation environment, in: *Principles of environmental physics.* pp. 23-38, London, Edward Arnold.
5. Zawilska, J.B., Berezińska, M., Rosiak, J., Skene, D.J., Vivien-Roels, B. & Nowak, J.Z. (2004) Suppression of melatonin biosynthesis in the chicken pineal gland by retinally perceived light – involvement of D1-dopamine receptors. *Journal of Pineal Research* 36, 80-86.
6. Morgan, I.G., Boelen, M.K. & Miethke, P. (1995) Parallel suppression of retinal and pineal melatonin synthesis by retinally mediated light. *Neuroreport* 6, 1530-1532.
7. Wight, P.A.L. (1971) The pineal gland, in: *Physiology and biochemistry of the domestic fowl* (Eds D.J. Bell & B.M. Freeman), pp. 549-573, London, Academic Press.
8. Takahashi, J.S. & Menaker, M. (1984) Multiple redundant circadian oscillators within the isolated avian pineal gland. Journal of Comparative Physiology [A] 154, 435-440.
9. Lauber, R.W., Boyd, J.E. & Axelrod, J. (1968) Enzymatic synthesis of melatonin in avian pineal body: Extraretinal response to light. *Science* 161, 489- 490.
10. Zawilska, J.B., Berezińska, M., Lorenc, A., Skene, D.J. & Nowak, J.Z. (2004) Retinal illumination phase shifts the circadian rhythm of serotonin N-acetyltransferase activity in the chicken pineal gland. *Neuroscience Letters* 360, 153-156.
11. Ciccone, N.A., Dunn, I.C., Boswell, T., Tsutsuni, K., Ubuka, T, Ukena, K & Sharp, P.J. (2004) Gonadotrophin inhibitory hormone depresses gonadotrophin α and follicle stimulating hormone β subunit expression in the pituitary of the domestic chicken. *Journal of Neuroendocrinology* 16, 999-1006.
12. Yoshimua, T., Yasuo, S., Watanabe, M., Iigo, M., Yamamura, T., Hirunagi, K. & Ebihara, S. (2003) Light-induced hormone conversion of T4 to T3 regulates photoperiodic response of gonads in birds. *Nature* 426, 178-181.
13. Proudman, J.A. & Siopes, T.D. (2002) Relative and absolute photorefractoriness in turkey hens: profiles of prolactin, thyroxine, and triiodothyronine early in the reproductive cycle. *Poultry Science* 81, 1218-1223.
14. Siopes, T.D. (1997) Transient hypothyroidism reinstates egg laying in turkey breeder hens: termination of photorefractoriness by propylthiouracil. *Poultry Science* 76, 1776-1782.
15. Lien, R.J. & Siopes, T.D. (1989) Effects of thyroidectomy on egg production, molt, and plasma thyroid hormone concentrations of turkey hens. *Poultry Science* 68, 1126-1132.
16. Lien, R.J. & Siopes, T.D. (1991) Influence of thyroidectomy on reproductive responses of male domestic turkeys (*Meleagris gallopavo*). *British Poultry Science* 32, 405-415.
17. Meddle, S.L. & Follett, B.K. (1997) Photoperiodically driven changes in Fos expression within the basal tuberal hypothalamus and median eminence of Japanese quail. *Journal of Neuroscience* 17, 8909-8918.
18. Siopes, T.D. & Wilson, W.O. (1980) Participation of the eyes in the photo-stimulation of chickens. *Poultry Science* 59, 1122-1125.
19. Benoit, J., Walter, F.X. and Assenmacher, I. (1950) Nouvelles recherches relatives à l'action de lumières de différentes longeurs d'onde sur la gonado-stimulation du canard male impubère. *Chronicles of the Royal Society of Biology* 144, 1206.
20. Benoit, J. (1964) The role of the eye and of the hypothalamus in the photo-stimulation of gonads in the duck. *Annals New York Academy of Science* 111, 204-216.

21. Foster, R.G. & Follett, B.K. (1985) The involvement of a rhodopsin-like photopigment in the photoperiodic response of the Japanese quail. *Journal of Comparative Physiology A*. 157, 519-528.
22. Hartwig, H.G. & van Veen, T. (1979) Spectral characteristics of visible radiation penetrating into the brain and stimulating extraretinal photo-receptors. *Journal of Comparative Physiology* 130, 277-282.
23. Benoit, J., Walter, F.X. and Assenmacher, I. (1950) Contribution à l'étude du réflexe optohypophysaire. Gonadostimulation chez le canard soumis à des radiations lumineuses de diverses longueurs d'onde. *Journal of Physiology* 42, 537-541.
24. Antonini, E. and Brunori, M. (1971) Haemoglobin and myoglobin in their reactions with ligands, in: *Frontiers of biology* (Eds A. Neuberger & E.L. Tatum) Vol. 21, pp. 13-54. Amsterdam, North Holland.
25. Mason, H.S., Ingram, D.J.E. & Allen, B. (1960) The free radical property of melanins. *Archives of Biochemistry and Biophysics,* **86**: 225-230.
26. Prayitno, D.S. & Phillips, C.J.C. (1997) Equating the perceived intensity of coloured lights to hens. *British Poultry Science* 38, 136-141.
27. Benoit, J. and Assenmacher, I. (1953) Action des facteurs externes et plus particulièrement du facteur lumineux sur l'activité sexualle des oiseaux, in: *Rapport à la IIème Réunion des Endocrinologistes de langue française*. pp. 33-80, Paris, Masson.
28. Pittendrigh, C.S. (1966) The circadian oscillator in *Drosophila pseudoobscura*: A model for the photoperiodic clock. *Zeitschrift fur Pflanzenphysiolie* 54, 275-307.
29. Hamner, W.M. (1963) Diurnal rhythm and photoperiodism in testicular recrudescence of the House finch. *Science* 142, 1294-1295.
30. Follett, B.K. & Sharp, P.J. (1969) Circadian rhythmicity in photoperiodically induced gonadotrophin release and gonadal growth in the quail. *Nature* 223, 968-971.
31. Nicholls, T.J., Follett, B.K. & Robinson, J.E. (1983) A photoperiodic response in gonadectomized Japanese quail exposed to a single long day. *Journal of Endocrinology* 97, 121-126.
32. Follett, B.K. (1981) The stimulation of luteinising hormone and follicle-stimulating hormone secretion in quail with complete and skeleton photoperiods. *General and Comparative Endocrinology* 45, 306-316.
33. Wilson, S.C. (1982) Evidence of a photoinducible phase for the release of luteinizing hormone in the domestic hen. *Journal of Endocrinology* 94, 397-406.
34. Siopes, T.D. and Wilson, W.O. (1980) A circadian rhythm in photosensitivity as the basis for the testicular response of Japanese quail to intermittent light. *Poultry Science* 59, 868-873.
35. Follett, B.K. & Milette, J.J. (1982) Photoperiodism in quail: testicular growth and maintenance under skeleton photoperiods. *Journal of Endocrinology* 93, 83-90.
36. Pittendrigh, C.S. & Minis, D.H. (1964) The entrainment of circadian oscillations by light and their role as photoperiodic clocks. *American Naturalist* 98, 261-294.
37. Bünning, E. (1936) Die endonome Tagesrhymik als Grundlage der photo-periodischen Reaktion. *Berichte der Deutschen Botanischen Gesellschaft* 54, 590-607.
38. Farner, D.S. (1964) The photoperiodic control of reproductive cycles of birds. *American Science* 52, 137-156.
39. Pittendrigh, C.S. (1981) Circadian organization and the photoperiodic phenomena, in: *Biological clocks in seasonal reproductive cycles* (Eds. B.K & D.E. Follett), pp. 1-35. Bristol, J. Wright & Sons.
40. Brandstatter, R. (2003) Encoding time of day and time of year by the avian circadian system. *Journal of Neuroendocrinology* 15, 398-404.
41. Meddle, S.L. & Follett, B.K. (1995) Photoperiodic activation of fos-like immunoreactive protein in neurones within the tuberal hypothalamus of Japanese quail. *Journal of Comparative Physiology* [A] 176, 78-79.

42. Yasuo, S., Watanabe, M., Okabayashi, N., Ebihara, S. & Yoshimura, T. (2003) Circadian clock genes and Photoperiodism: Comprehensive analysis of clock gene expression in the mediobasal hypothalamus, the suprachiasmatic nucleus, and the pineal gland of Japanese quail under various light schedules. *Endocrinology* 144, 3742-3748.
43. Farner, D.S., Mewaldt, L.R. & Irving, S.D. (1953) The roles of darkness and light in the activation of avian gonads. *Science* 118, 351-352.
44. Follett, B.K., Robinson, J.E., Simpson, S.M. & Harlow, C.R. (1981) Photoperiodic time measurement and gonadotrophin secretion in quail, in: *Biological clocks in seasonal reproductive cycles.* (Eds. B.K & D.E. Follett), pp. 185-201. Bristol, J. Wright & Sons.
45. Follett, B.K., Farner, D.S. & Morton, M.L. (1967) The effects of alternating long and short daily photoperiods on gonadal growth and pituitary gonadotrophins in the White crowned sparrow. *Biological Bulletin* 133, 330-342.
46. Farner, D.S., Downham, R.S., Lewis, R.A., Mattocks, R.W., Darden, T.R. & Smith, J.P. (1977) The circadian component in the photoperiodic mechanism of the House sparrow, *Passer domesticus*. *Physiological Zoology* 50, 247-268.
47. Lewis, P.D., Perry, G.C. & Morris, T.R. (1997) Responses of immature pullets to repeated cycles of gradual increases and abrupt decreases in photoperiod. *British Poultry Science* 38, 611-613.
48. Lewis, P.D. & Perry, G.C. (1988) Responses of the laying hen to irregular illumination. *British Poultry Science* 29, 875-876.
49. Dawson, A., King, V.M., Bentley, G.E. & Ball, G.F. (2001) Photoperiodic control of seasonality in birds. *Journal of Biological Rhythms* 16, 365-380.
50. Siopes, T.D. (1984) Recycling turkey hens with low light intensity. *Poultry Science* 63, 1449-1452.
51. Woodard, A.E., Hermes, J.C. & Fuqua, C.L. (1986) Effects of light conditioning on reproduction in partridge. *Poultry Science* 65, 2015-2022.
52. Dawson, A. & Goldsmith, A.R. (1983) Plasma prolactin and gonadotrophins during gonadal development and the onset of photorefractoriness in male and female starlings (Sturnus vulgaris) on artificial photoperiods. *Journal of Endocrinology* 97, 253-260.
53. Woodard, A.E., Abplanalp, H. & Snyder, R.L. (1980) Photorefractoriness and sexual response in aging partridge kept under constant long- and short-day photoperiods. *Poultry Science* 59, 2145-2150.
54. Lewis, P.D., Backhouse, D. & Gous, R.M. (2004) Constant photoperiods and sexual maturity in broiler breeder pullets. *British Poultry Science* 45, 557-560.
55. Gous, R.M. & Cherry, P. (2004) Effects of body weight at, and lighting regimen and growth curve to, 20 weeks on laying performance in broiler breeders. *British Poultry Science* 45, 445-452.
56. Lewis, P.D., Ciacciariello, M. & Gous, R.M. (2003) Photorefractoriness in broiler breeders: sexual maturity and egg production evidence. *British Poultry Science* 44, 634-642.
57. Follett, B.K. (1991) The physiology of puberty in seasonally breeding birds, in: *Follicle stimulating hormone: regulation of secretion and molecular mechanisms of action* (Eds. M. Hunzicker-Dunn & N.B. Schartz), pp. 54-65. New York, Springer-Verlag.
58. Farner, D.S. & Follett, B.K. (1966) Light and other environmental factors affecting avian reproduction. *Journal of Animal Science* 25 (supplement), 90-115.
59. Robinson, J.E. & Follett, B.K. (1982) Photoperiodism in Japanese quail: the termination of seasonal breeding by photorefractoriness. *Proceedings of the Royal Society, London* 215, 95-116.
60. Siopes, T.D. (2002) Photorefractoriness in turkey breeder hens is affected by age and season. *Poultry Science* 81, 689-694.
61. Siopes, T.D. (2001) Temporal characteristics and incidence of photo-refractoriness in turkey hens. *Poultry Science* 80, 95-100.
62. Siopes, T.D. (2005) Spontaneous recovery of photosensitivity by turkey breeder hens given prolonged exposure to long day lengths. *Poultry Science* 84, 1470-1476.
63. Lewis, P.D. & Morris, T.R. (2005) Change in the effect of constant photoperiods on the rate of sexual maturation in modern genotypes of domestic pullet. *British Poultry Science* 46, 584-586.

64. Morris, T.R., Sharp, P.J. & Butler, E.A. (1995) A test for photorefractoriness in high-producing stocks of laying pullets. *British Poultry Science* 36, 763-769.
65. Rosiak, J. & Zawilska, J.B. (2005) Near-violet light perceived by the retina generates the signal suppressing melatonin synthesis in the chick pineal gland - an involvement of NMDA glutamate receptors. *Neuroscience Letters* 379, 214-217.

3 PHOTOPERIOD
Conventional programmes

Photoperiod is synonymous with daylength in most contexts, but we sometimes also use the term 'daylength' to describe the period of rotation of a planet, which here on earth is 24 h. A conventional lighting programme cycles every 24 h, and contains a single photoperiod that is interpreted as day and a single period of darkness (called the scotoperiod) that is interpreted as night. Programmes that do not cycle every 24 h, but still contain only one photoperiod and one scotoperiod, are called ahemeral cycles. They may be shorter or longer than 24 h.

Although photoperiods are often classified as stimulatory (long day) or non-stimulatory (short day), poultry respond most dramatically to changes in photoperiod. Photoperiodic responses vary with species and type of stock. In this chapter, responses of each class of poultry (growing pullets, laying hens, broiler parents, broilers, turkeys and waterfowl) are described, first for constant photoperiods and, secondly, for changing photoperiods.

Ahemeral cycles, intermittent lighting, and continuous lighting are discussed in Chapter 4.

• Growing pullets: effects of constant photoperiods •

The main effect of photoperiod during the rearing phase is its control over the timing of sexual maturation. The response to constant photoperiods was first described by a curvilinear model, which predicted that earliest maturity would occur when pullets were maintained on 17-h daylengths (1).

However, this was subsequently replaced by a model using two linear regressions either side of a hinge-point (Figure 3.1). This predicted that the earliest mean age at first egg (AFE) would be achieved by rearing birds on 10-h photoperiods (2). The model indicated that, for daylengths shorter than 10 h, AFE would occur 1.73 d later for each 1-h reduction in photoperiod, but that it would also be delayed by 0.30 d for each hour that daylength exceeded 10 h. It was acknowledged that the reduction in feed intake and suppression of growth that occurs when pullets are reared on short days might be partly responsible for the delay in sexual maturation. In modern hybrids, this effect on feed intake during the rearing period has been reported to be about -4% for each 1-h reduction in photoperiod (3). The effect of lower feed intakes on AFE has been estimated as 0.64 d (4) or 0.53 d (5) for each 1% reduction in consumption. However, the findings of an experiment in which pullets were given *ad libitum* feed in mash or pellet form showed that shorter daylengths did not retard AFE through limitation of nutrient intake (6). Photoperiodically induced reductions in feed intake are insufficient to account for the observed delay in AFE on very short days, while pullets on daylengths longer than 10 h mature later despite eating more feed (7). We can conclude that short days exert a direct retarding influence on the hypothalamic-pituitary system and are not neutral, as has previously been proposed (8), but are simply less stimulatory to gonadotrophin releasing hormone (GnRH) and gonadotrophin release than a 10-h day. The reasons for the small but significant increasing delay in AFE as daylength is extended beyond 10 h are far from clear. Juvenile photorefracroriness, which dissipates more slowly on long days, has been suggested as an explanation, but this theory does not hold when predictions are made for the response of egg-type pullets to changes in daylength.

The model for the effect of constant photoperiod on AFE was further refined when data for modern (post-1993) egg-type hybrids showed that they respond more strongly to photoperiods < 10 h than did earlier (pre 1968) hybrids (Figure 3.1) (9). The new model predicts:

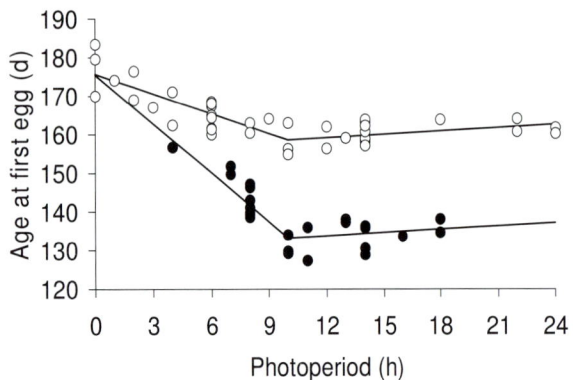

Figure 3.1 *Mean age at first egg (d) for ad libitum-fed early (○) and modern (●) hybrids maintained on constant photoperiods (9).*

for $p \leq 10$ h   AFE (d) = 175.5 − 4.222p
for $p \geq 10$ h   AFE (d) = 130.4 + 0.285p

where p = photoperiod (h).

Thus the retarding effect on AFE in modern birds is 4.2 d for each 1-h reduction in daylength below 10 h, more than double the rate for the pre-1968 hybrids. However, there is no noticeable change in the rate of response to photoperiods that are longer than 10 h. The likely reason for this is that breeding companies have advanced mean AFE by 3 to 4 weeks, as a result of selection for egg numbers, but that this genetic advantage is only expressed in pullets that are maintained on daylengths similar to those on which the birds were selected, which is 8 to 10 h. A parallel phenomenon has been observed in broilers. Modern stocks only show their full superiority over earlier meat-strains when given diets formulated to modern nutritional specifications (10).

It is common commercial practice to provide growing pullets with a short period of step-down lighting before growing them on a constant short-day. This is to ensure that the chicks have explored their environment and started eating and drinking satisfactorily. Maturity will be delayed by 1-2 d for each week taken to reach the constant short-day, depending on the stimulatoriness of the subsequent lighting regimen; the less stimulatory the programme the bigger the retarding effect of the step-down lighting (11). Whilst this initial phase of longer daylengths results in higher feed intakes and faster initial growth, it does not give any improvement in total egg yield, simply fewer but larger eggs as a result of the delay in AFE (11, 12).

Feed intake

Cumulative feed intake to 18 weeks increases curvilinearly with photoperiod (Figure 3.2). The regression, with differences between intakes for white-egg and brown-egg hybrids removed by least squares (white-egg hybrids -1.12 kg), is described as:

$$y = 7.61 - 0.160p + 0.0144p^2$$

where y = feed intake to 18 weeks (kg) and p = photoperiod (h). It is likely that the similarity of intake below 8 h is a consequence of pullets on very short daylengths learning to eat in the dark. This same phenomenon also occurs in broilers (page 38) and growing turkeys (page 40).

However, the quantity of feed required to reach sexual maturity is dependent on both the photoperiod and on AFE as influenced by the photoperiod. Mean consumption to first egg for brown- and white-egg hybrids is described by the following equation (3):

$$FI = -11.98 + 0.221p + 0.133 AFE$$

where FI = cumulative feed intake to first egg (kg), p = photoperiod (h) and AFE = mean age at first egg (d).

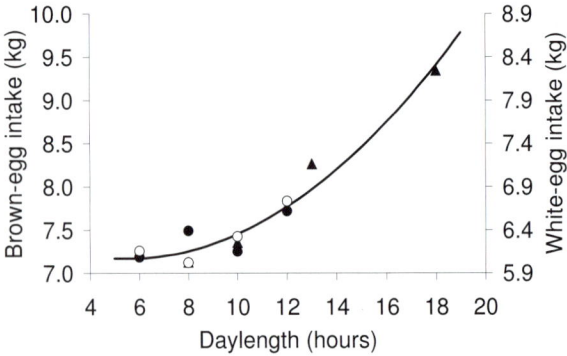

Figure 3.2 *Cumulative feed intake to 18 weeks of age for brown-egg (●), white-egg (○) and mixed brown-egg and white-egg (▲) hybrids. (3 and unpublished data from the University of Guelph).*

• Growing pullets: effects of changing photoperiod •

Growing pullets respond more to a change in photoperiod than to the initial or final photoperiods themselves, irrespective of whether the change is an increase or a decrease (1). Data from a trial in which white-egg pullets were maintained on 8, 13 or 18-h photoperiods or transferred from 8 to 13 h, 13 to 18 h, 18 to 13 h or 13 to 8 h at 12 weeks provide a good example (7). Figure 3.3 shows that whereas there was only a 3-d difference among the AFE means for pullets given constant daylengths, more than 43 d separated the earliest treatment (an increase from 8 to 13 h) from the latest maturing group (a decrease from 13 to 8 h).

These data also demonstrate that the influences of the initial and final photoperiod are more powerful than the size of the change. Although both the increments in this trial were of 5 h, the transfer from 8 to 13 h advanced AFE by 22 d whilst the 5-h increment from 13 to 18 h advanced mean AFE by less than a day. Likewise, the decrease from 18 to 13 h delayed maturity by only 1 d, whilst the same sized decrease from 13 to 8 h delayed AFE by 23 d, all differences expressed relative to pullets maintained on the initial photoperiod. This shows that the effect of a change in daylength depends on the stimulatoriness of the two photoperiods. A change between two stimulatory or between two non-stimulatory photoperiods has a lesser effect on AFE than a change between a stimulatory and a non-stimulatory photoperiod, irrespective of the direction of change.

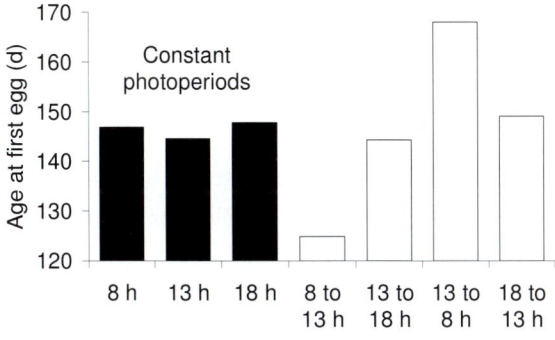

Figure 3.3 *Mean age at first egg (d) for pullets maintained on 8, 13 or 18 h photoperiods (solid) or given a 5-h increase or decrease (open) at 12 weeks (7).*

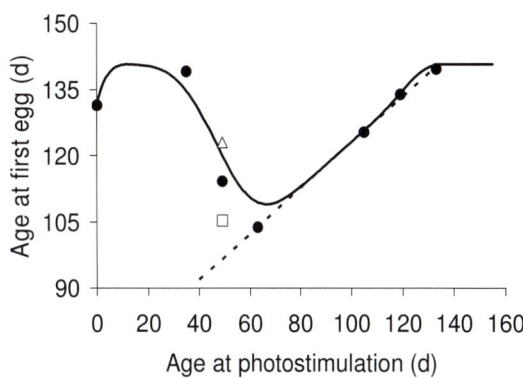

Figure 3.4 *Mean age at first egg for pullets transferred from 8 to 16-h photoperiods at various ages and the fitted model. The mean for constant 16-h controls is plotted at 0 d. The mean at 49 d is a mixture of responders (△) and non-responders (□). The dotted line describes the linear regression between 9 and 19 weeks (13).*

The other important factor affecting the timing of sexual maturation is the age at which the changes in daylength are given (13). A flock of modern full-fed pullets do not respond to an increment in photoperiod until about 5 weeks of age, and not all birds within the flock are photo-responsive to an increase until about 9 weeks (Figure 3.4). The reason why increments in photoperiod at very young ages do not advance AFE, despite being able to induce an increase in plasma luteinizing hormone (LH) concentration (14), is that the hypothalamic-pituitary axis is insufficiently developed to stimulate the release of follicle stimulating hormone (15); a consequence, in part, of suboptimal plasma concentrations of oestrogen (16, 17).

The proportion of birds that have become photosensitive at a given age between 5 and 9 weeks forms a normal distribution with a mean of 50 d and a standard deviation of 7.4 d.

From 9 weeks until about 2 weeks before the mean age at which the birds would have started maturing spontaneously in response to the initial photoperiod, the effect of an increase in daylength progressively decreases at a rate that is dependent on the initial and final photoperiod, and the genotype.

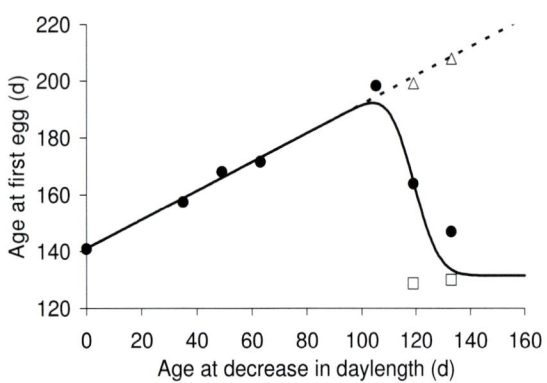

Figure 3.5 *Mean age at first egg for pullets transferred from 16 to 8-h photoperiods at various ages (●) and the fitted model. Means at 119 and 133 d are a mixture of responders (△) and non-responders (□). The dotted line describes the linear regression between 0 and 19 weeks (13).*

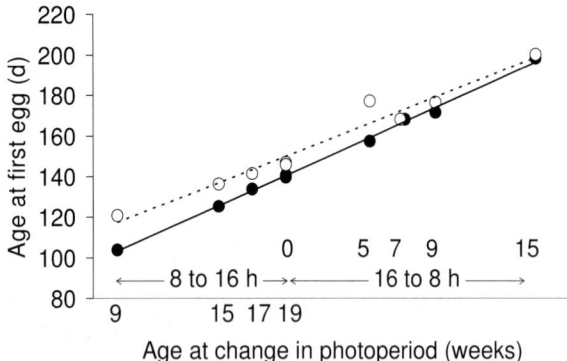

Figure 3.6 *Effect of age at transfer on AFE (d) for brown-egg (●) and white-egg (○) pullets changed from 8 to 16 h or from 16 to 8 h photoperiods (13).*

In contrast, a flock of pullets will respond to a decrease in daylength from day-old and continue to do so until the first birds start maturing spontaneously at about 15 weeks. The magnitude of the response increases with age (Figure 3.5). After 15 weeks, as with the attainment of photoresponsiveness, the proportion of birds that can no longer postpone their first egg forms a normal distribution with a mean equal to that of the predicted AFE (A) and a standard deviation calculated as SD = $0.0623A$.

The rate of change in AFE with age (that is, the slopes of the lines illustrated in Figures 3.4 and 3.5) depends on both initial and final photoperiod, and is predicted by the following equation:

$$b = k_i(0.1338 + 0.1496C - 0.01884C^2 + 0.0009683C^3 - 0.00001941C^4 - 0.22396M + 0.05028M^2 - 0.00365M^3 + 0.00008216M^4)$$

where b = the change in AFE (d) for each 1-d delay in applying the change in photoperiod, C = difference between the initial and final photoperiod (h), M = mean of the initial and final photoperiods (h) and k_i = an adjustment for the difference in responsiveness between the given genotype and ISA Brown hybrids. The b value is the same for any pair of photoperiods, irrespective of the direction of change (Figure 3.6). For example, b = 0.472 for a 6 to 12 h and for a 12 to 6 h transfer. The largest predicted b value (0.508), that is the combination of photoperiods that induces the biggest advance or delay in AFE, is for a transfer between 7 and 14 h.

The retarding effect of a decrease in daylength continues until about 15 weeks, but the advancing influence of an increment, because it is not fully effective until about 9 weeks, lasts for only half this time, and, as a consequence, decreases in daylength have double the potential to delay AFE that increments have to advance it.

All of these photoperiodic influences have been included in a model that predicts AFE for *ad libitum*-fed pullets given a single change in photoperiod (13). The equation for an increase in photoperiod is:

$$A_m = (1-p)A + p(1-m)[A-b(A-t)] + pmA$$

and for a decrease in photoperiod:

$$A_m = (1-m)(A+bt) + mA$$

where A_m = predicted mean age at first egg (d), A = mean AFE (d) for pullets of a particular genotype maintained on the initial photoperiod, p = proportion of birds that have become sensitive to an increment in photoperiod, m = proportion of birds that have spontaneously started rapid gonadal development in response to the initial photoperiod and are within 10 d of laying their first egg, b = the rate of change in

AFE (d) for a 1-d delay in transferring the pullets to the final photoperiod and t = the age (d) at which the increase or decrease in photoperiod is given.

Two changes in photoperiod

In general terms, when a pullet experiences two changes in photoperiod, as commonly occurs in a commercial situation, the first change alters its physiological age so that it responds to the second change as if it had been given at this amended age rather than at its chronological age (18). For example, if the initial change was a decrease in daylength that would have retarded AFE by 20 d, the pullet will be physiologically 20 d younger and will respond to an increase given at 105 d as if it had been applied at 85 d, and AFE will be advanced to a greater degree than if this had been the only change in daylength. Similarly, if the initial change is an increment that would have advanced AFE, the pullet will respond to a subsequent decrease in photoperiod as if it had been given at a later age and the delay in AFE will be greater than if there had been no initial increase. However, there will be a lag following the first change before the physiological age of the pullet will be changed, and so, if the second change takes place soon after the first, the pullet will respond to it as if it was the only change, i.e. at its chronological age. Where an increase in daylength is given to photosensitive pullets and followed shortly afterwards by a reciprocal decrease, there will be no change in AFE. This means that temporary transfers to long days can be used for management purposes, for example, to permit feeding and drinking after pullets have had a long journey during the transfer from the rearing to the laying facilities, without affecting sexual maturity (19). Unpublished data from the University of KwaZulu-Natal show that whilst 8 d of 14-h photoperiods are needed to advance maturity in the earliest maturing birds in a flock, 20 d are probably needed to advance AFE in all birds when they are transferred from 8-h days.

Feed intake

Feed intake increases following an increment in photoperiod due to the greater feeding opportunity and to an increased energy demand (birds expend 1.3% more energy in a 1-h photoperiod than in 1 h of darkness). This increase in nutrient intake then makes a small contribution to the advance in AFE by accelerating growth and fat deposition. For example, whereas AFE in *ad libitum*-fed brown-egg hybrid pullets was advanced by 35 d by a transfer from 8 to 13-h photoperiods at 9 weeks, AFE was 31 d earlier when feed intake was limited to that of pullets maintained on 8-h photoperiods (20). The 4-d later AFE was associated with an 8% lower feed intake, and this agrees remarkably well with the previously reported effect of a 0.53 d delay in sexual maturity for each 1% reduction in feed intake (5).

Natural light

Where birds are housed in non-controlled environment facilities, they are obviously influenced by naturally changing daylengths. The effect that these changes have on AFE will vary with time of year and the latitude-dependent difference between the shortest and longest day. This seasonal effect, first reported in 1918 (21), was subsequently quantified in an extensive experiment in which pullets were hatched at weekly intervals over a 21-month period (22). Natural daylength (sunrise to sunset) at the research farm (51°27'N) varied from 7.8 h on the shortest day to 16.6 h on the longest day. There was a range of 33 d in age at sexual maturity (Figure 3.7), with the earliest maturity occurring in January-hatched pullets and the latest maturity in June-hatched pullets.

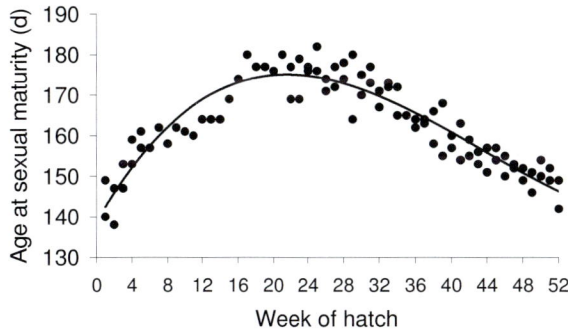

Figure 3.7 *Age at first egg (AFE) for pullets hatched at weekly intervals over a 21-month period; January = 0 (22). The prediction equation is (x = week of hatch):*

$$AFE = 138.8 + 3.79x - 0.117x^2 + 0.00091x^3$$

Chapter 3 Photoperiod: Conventional programmes

• Laying hens: effects of constant photoperiods •

Egg numbers

The annual rate of egg production by laying hens that have been kept on a constant photoperiod from hatch or given a single increment before they become sexually mature, increases by about 4 eggs for each 1-h increase in daylength up to 10 h. Beyond 10 h, the response to photoperiod depends on genotype, with rate of lay levelling off in white-egg hybrids but continuing to increase in brown-egg hybrids up to 13 or 14 h (Figure 3.8). Beyond 14 h, there is no further improvement in egg production in either type of stock, and, indeed, it might even decrease in hens given daylengths longer than 17 h (3).

Egg weight

Mean egg weight over the laying year for birds given a single increase in daylength at the end of the rearing period increases linearly with photoperiod by 0.1 to 0.2 g per 1 h in both white and brown-egg hybrids (3). However, the rate of increase tends to be slightly higher in white-egg layers than in brown-egg layers due, perhaps, to the levelling of egg production above 10 h in the white-egg birds. The result of these dissimilar responses by white and brown-egg hybrids is that mean daily egg mass output increases by 0.5 g/h on average over the range 8 to 18-h photoperiods in brown-egg hybrids, but by only 0.3 g/h in white-egg hens.

Shell quality

Shell weight and shell thickness index (mg/cm^2) decrease linearly with increasing photoperiod (Figure 3.9, 23). However, the rate of deterioration is greater in brown-egg than white-egg hybrids, possibly as a consequence of their higher egg mass output (about 2 g/d on 14 h daylengths). Shell quality is better in hens given shorter daylengths, but it is probably the longer scotoperiod that produces this effect rather than the shorter photoperiod. This is because the peak concentrations of calcitonin (24) and parathyroid hormone (25), the two hormones involved in the mobilisation of calcium for the shell, occur at night. Evidence from rats for the effect of lighting schedule on the diurnal rhythm of bone resorption suggests that parathyroid hormone synthesis is closely linked to the concentration of plasma melatonin (26), the nocturnally synthesised hormone implicated in photoperiodic time measurement (page 15). Melatonin rhythmicity has also been reported to be associated with modifications of osteoclast and osteoblast activity (26, 27). It is reasonable to conclude that short photoperiods and long scotoperiods permit the mineral mobilisation hormones to influence the release of calcium from the skeleton over a longer period than do long days and short nights, and that this extended activity results in thicker shells. The incidence of body-checked eggs, also termed

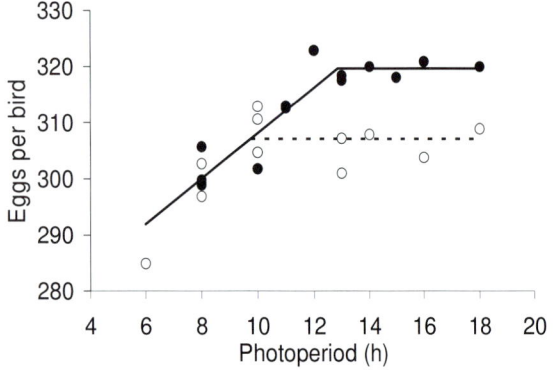

Figure 3.8 *Annual egg production for brown-egg (●) and white-egg (○) hybrids given constant photoperiods from hatch or a single increment prior to sexual maturity (3).*

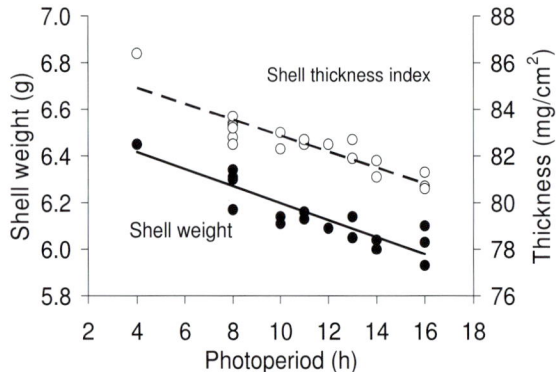

Figure 3.9 *Mean shell weight (g) and shell thickness index (mg/cm^2) in hens given constant daylengths from hatch or a single increment prior to sexual maturity (23).*

equatorial bulges, increases with photoperiod (28). This condition is caused by an adrenalin-induced involuntary contraction of uterine muscles when the egg is at a fragile stage of shell mineralisation. The uterus later 'papers over' the cracks, and this creates the bulge around the equator of the egg. The stress that induces the adrenalin surge is temporally associated with the end of the photoperiod; so the proportion of body-checked eggs is higher in the morning than at other times of the day. Whilst the incidence of body-checks has been reduced experimentally by shortening the daylength (28), this remedial action cannot be taken in practice because it is likely to result in a reduction in rate of lay. However, the provision of a 15-min dusk period may help to reduce stress, especially in birds kept in cages or aviary systems, by allowing them to settle for the night in a more orderly fashion (discussed later in the dawn and dusk section on page 48). The provision of a period of twilight is recommended in European Union (EU) laying hen welfare regulations, especially for birds kept in alternative systems and enriched cages (29).

Feed intake

Daily feed intake is positively affected by photoperiod, increasing by about 1.3 g for each 1-h extension of the photoperiod in modern egg-laying hybrids (Figure 3.10, 3).

Whilst increased feeding opportunity may make a contribution towards this larger appetite, the main influencing factor is an increase of 1.3% in energy expenditure for each extra hour of illumination. In brown-egg hybrids, this has been shown to induce an increase of 9.3 kJ/kgW$^{0.75}$.h in daily metabolizable energy intake, which is reflected in a 7.8 kJ/kgW$^{0.75}$ ($\approx$1%) rise in heat production (30). This latter figure agrees well with predictions for energy expenditure made from calorimetric measurements of diurnal variations in heat production (31). Although laying hens given short photoperiods tend to start eating before the beginning of the photoperiod and to have a higher initial rate of ingestion, feeding activity during the final 6 h of the photoperiod is similar under 8, 11 and 14-h photoperiods (32), perhaps because the hens have a common demand for calcium for egg formation (33) or they simply anticipate nocturnal fasting (34).

Feed conversion efficiency

The combined effect of the photoperiodic influences on sexual maturity, rate of lay, egg weight and feed intake is that, for photoperiods above 10 h, egg output per unit of feed intake deteriorates linearly by 0.0035 g/g of feed intake for each 1-h increment in photoperiod (Figure 3.11, 3).

Mortality

Mortality rates during the laying year increase with longer photoperiods. Meta-analysis has shown that mortality for a brown-egg hybrid rises from 2.8 to 8.3% and from 5.6 to 13.9% for a white-egg hybrid as daylength is increased from 8 to 16 h (35). One contributing factor is an over-consumption of energy on longer days,

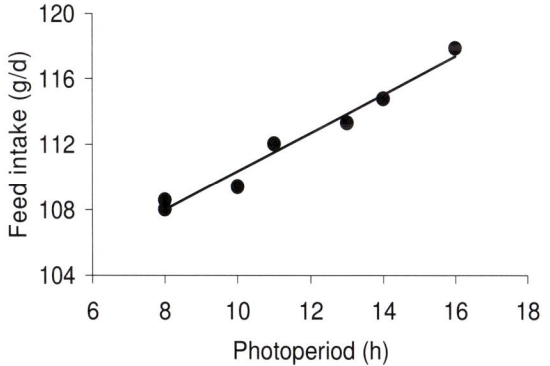

Figure 3.10 *Mean daily feed intake (g) for brown-egg and white-egg hybrids given constant photoperiods from hatch or a single increment prior to sexual maturity (3).*

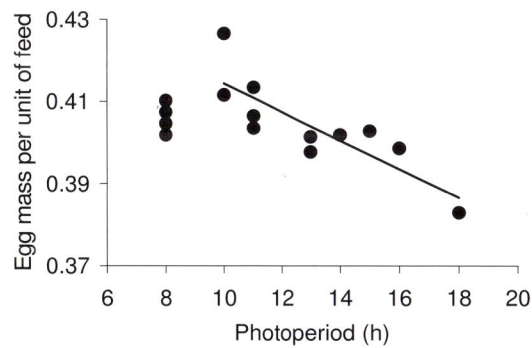

Figure 3.11 *Mean egg output per unit of feed intake for laying hens given constant photoperiods from hatch or a single increment prior to sexual maturity (3).*

which predisposes to an increase in the incidence of fatty liver haemorrhage.

Additionally, longer daylengths may exacerbate behavioural vices such as feather pecking with deleterious consequences for survival.

Time of lay

The surge of LH which initiates ovulation is limited to an 8 to 10-h period in the night, termed the open-period (36). This period, in the case of the fowl, is located centrally within the night. It is primarily positioned by dusk, but is also influenced by changes in the timing of dawn (37). The effect of photoperiod on mean oviposition time (MOT) varies with genotype and degree of selection for rate of lay. For the same photoperiod, brown-egg hybrids tend to lay 1.2 to 1.4 h earlier in the day than white-egg hybrids (38). Brown-egg hybrids exposed to 18-h photoperiods have been reported to start daily egg-laying more abruptly than birds given 8, 10 or 13-h photoperiods (38). This may be due to ovulations for the first eggs of the day being delayed because low plasma concentrations of melatonin towards the end of the photoperiod prevent the necessary pre-ovulatory surges in LH (39). The continuing intensive selection for rate of lay in modern egg-type hybrids has resulted in not only genetically better egg producing birds with early ages at first egg, but also in birds that form an egg more quickly (40). The location of the open-period for a given photoperiod will be fixed, and so a shorter egg formation time results in an earlier MOT; this is advancing by about 10 min per generation (41). The effect of photoperiod on mean oviposition time is shown in Figure 3.12, with the data adjusted by least squares to that for brown-egg hybrids laying in 1995. MOT for photoperiods (*p*) between 1 and 23 h is best described by a cubic regression:

$$MOT = -7.55 + 1.51p - 0.102p^2 + 0.0033p^3$$
$$(r^2=0.972, P<0.001, SD=0.75)$$

where *MOT* = mean oviposition time relative to dawn (h) and *p* = photoperiod (h).

However, over a more practical range of photoperiods from 8 to 18 h (● in Figure 3.12), a linear regression is as good a fit as any. MOT is delayed by about 30 min for each 1-h extension of the photoperiod:

$$MOT = -4.36 + 0.51p$$
$$(r^2=0.907, P<0.001, SD=0.73)$$

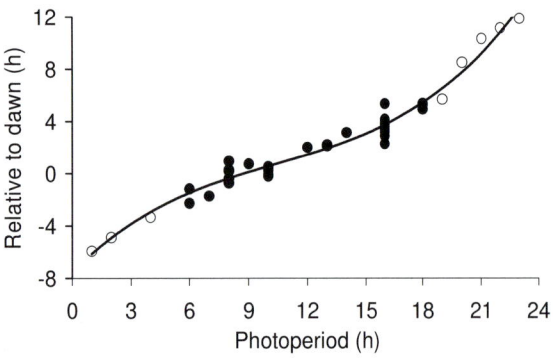

Figure 3.12 *Effect of photoperiod on mean oviposition time, relative to dawn, for egg-type hybrids (data for white-egg strains adjusted to modern brown-egg hybrids by least squares) (37, 38).*

• Laying hens: effects of changing photoperiods •

It is more common for pullets to be given a series of increments in photoperiod after they have been transferred to the laying facilities than to be given a single abrupt change in photoperiod. This practice stems from early days when lighting programmes mimicked natural changes in daylength (42). However, the use of step-up regimes appears to reflect the traditions of the poultry producer more than the biological needs of the pullet. The initial increase in daylength given to modern egg-type hybrids has two principle aims; to stimulate feed intake and to initiate rapid gonadal development. It is not uncommon for a flock of laying hens to suffer a post-peak dip in egg production. One possible reason is that their initial feed intake was insufficient to maintain body weight gain whilst growing the reproductive organs. It follows, therefore, that even if the birds are subsequently given further smaller increments, the first increase in photoperiod should be large enough to encourage an increase in appetite. It

Chapter 3 Photoperiod: Conventional programmes

also needs to exceed the saturation daylength for GnRH and gonadotrophin release if one wishes to optimise the photosexual response and, as a result, maximise peak rate of lay (43). Figure 3.13 suggests that saturation daylength is between 12.8 and 15.3 h for brown-egg pullets reared on 8 h (14) and about 14 h for white-egg pullets reared on 6 h (43). These data correlate well with the prediction that the maximum advance in AFE for a given photostimulation age is achieved by a transfer from 7 to 14-h photoperiods (13). Notwithstanding that abrupt increases to stimulatory photoperiods induce slightly earlier sexual maturity and result in higher initial rates of lay, the total egg output to the end of the laying year is not significantly different if hens are given a series of smaller increments in photoperiod (11).

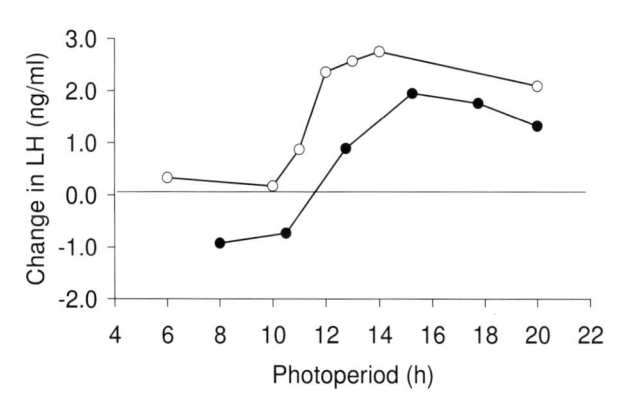

Figure 3.13 *Change in plasma LH 5 d after a transfer from 6 h for white-egg hybrids (○) or 8 h for brown-egg hybrids (●) to various photoperiods (14, 43).*

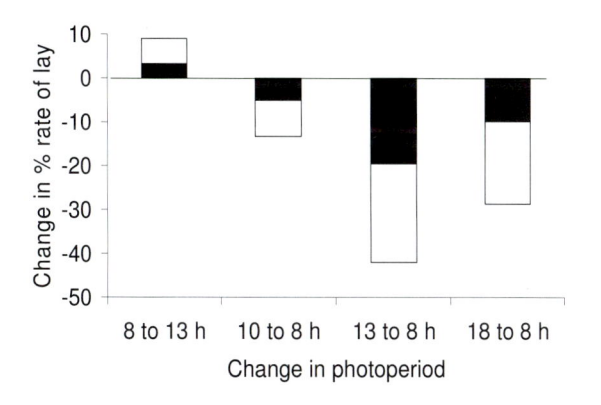

Figure 3.14 *The effect of a change in photoperiod at 31 weeks of age, relative to unchanged controls, on rate of lay in brown-egg and white-egg hybrids partitioned into the endocrine (solid) and nutritional (white) responses (44).*

Mature hens also respond to changes in daylength, with rate of lay being affected by influences on gonadotrophin release and nutrient intake (Figure 3.14). Data for modern hybrids show that rate of lay is linearly related to feeding opportunity but curvilinearly related to change in photoperiod (44). The consequence of these differences in response is that the relative importance of each influencing factor will vary with the size and direction of a daylength change. The non-linear photoperiodic response occurs because changes in LH and FSH output depend, not on the size of the light change, but on the particular photoperiods involved. The 10-h change from 18 to 8 h in this trial had only half the effect of the 5-h change from 13 to 8 h. These findings are in contrast to results with earlier hybrids, where feeding opportunity played no part in determining the response to a change in daylength (45). However, the modern birds received their diet in mash, as opposed to pelleted form, and had much smaller appetites, which probably explains the disparity. Nevertheless, the effects on rate of lay were similar, with increases in daylength enhancing and reductions depressing performance. When rate of lay is in excess of 90%, increases in photoperiod necessarily have less opportunity to enhance production than a decrease has to suppress it, and the findings probably do not indicate that the modern bird responds differently to photoperiodic increases and decreases.

Daily feed intake for brown- and white-egg hybrids altered by 2.5 g, relative to controls, for each 1-h difference in daylength (Figure 3.15).

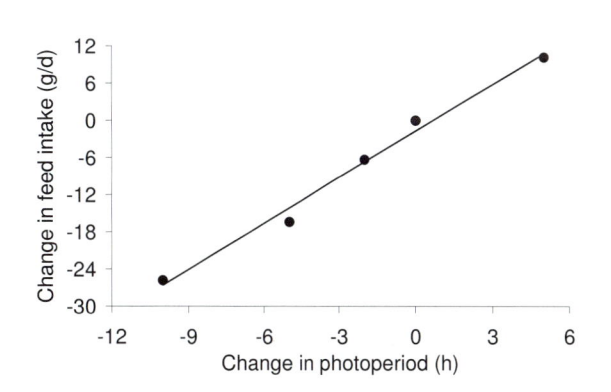

Figure 3.15 *Change in feed intake relative to constant-day controls following a change in photoperiod (44).*

Egg weight

Effects on egg weight are generally small but, as might be expected, are in the opposite direction to those for rate of lay. These and other data (46) suggest that a change in photoperiod evokes endocrine changes that control the rate of ovulation, that alterations in feeding opportunity influence nutrient intake to determine egg output, and that the resulting egg weight is a consequence of these two factors. Changes in feed intake influenced by changes in feeding opportunity when daylength is extended or reduced also affect body weight. In the experiment that used modern hybrids, egg weight increased by 0.2 g/d for each 1-h increment in daylength (44).

As laying hens age, rate of lay reduces in association with reduced plasma and pituitary LH concentrations, and a lowered responsiveness to GnRH (47). It has been suggested that once birds have started egg production, the aim should be to increase the photoperiod just sufficiently to balance any development of photorefractoriness (43). However, with modern hybrids capable of laying in excess of 330 eggs in a 12-month cycle, there seems little prospect of such a lighting programme enhancing performance. Indeed, the conclusion from an experiment that tested such a regimen was that it gave no advantage to egg numbers, egg weight, egg output, feed intake or feed conversion efficiency (48).

• Broiler breeders: effects of constant photoperiods •

Broiler breeders are not fed *ad libitum*, and respond differently to photoperiodic treatments when they are managed to achieve different growth profiles. However, when birds are given constant photoperiods, the response appears to be specific to daylength and lower body weight simply delays maturity for all treatments (Figure 3.16, 49). Broiler breeding stock exhibit photorefractoriness, and so sexual maturity in response to constant non-stimulatory daylengths (< 10 h) is similar to that for egg-type birds but markedly different for birds exposed to longer daylengths (> 10 h). Broiler breeders are hatched in a state of juvenile photorefractoriness, and the rate of its dissipation appears to be proportional to the potential stimulatoriness of the photoperiod and not to its length *per se*, as reported for exotic avian species (50). Data, adjusted by least squares analysis for differences in body weight and among experiments, show that the earliest sexual maturity is achieved by providing 10-h daylengths and the latest when birds are reared on constant 13 or 14 h (Figure 3.17).

The effect of constant photoperiods on *ad libitum* feed intake has not been determined because this is irrelevant for broiler breeders allocated fixed quantities of feed. However, photoperiod does have an effect on body weight gain despite birds being given the same amounts

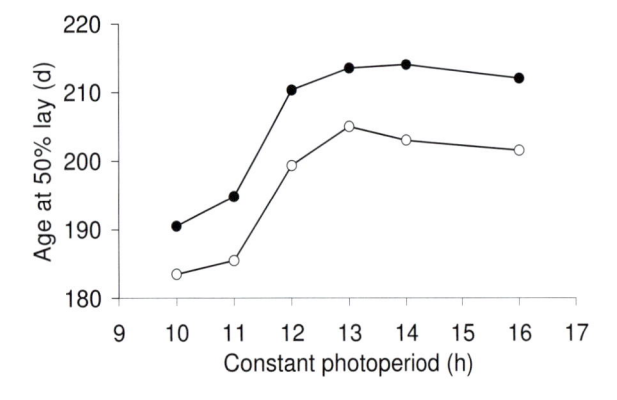

Figure 3.16 *Mean age at 50% lay for broiler breeders grown to reach 2.1 kg body weight at 17 (○) or 21 (●) weeks of age and maintained from 2 d of age on 10, 11, 12, 13, 14 or 16-h photoperiods (49).*

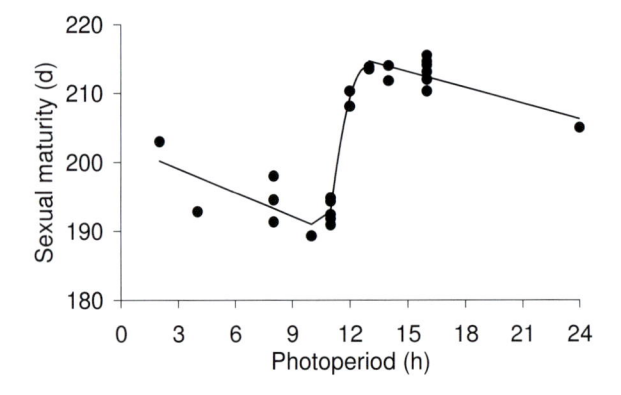

Figure 3.17 *Mean age at sexual maturity for broiler breeders maintained on various constant photoperiods, adjusted by least squares to the response of birds reaching 2.1 kg body weight at 20 weeks (49, 51, 52).*

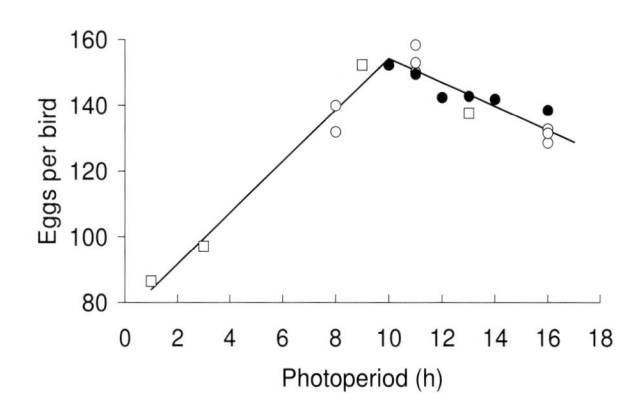

Figure 3.18 *Total egg production to 60 weeks of age for broiler breeders maintained on various photoperiods, with differences among trials removed by least squares analysis (○ 52, ● 53, □ 55).*

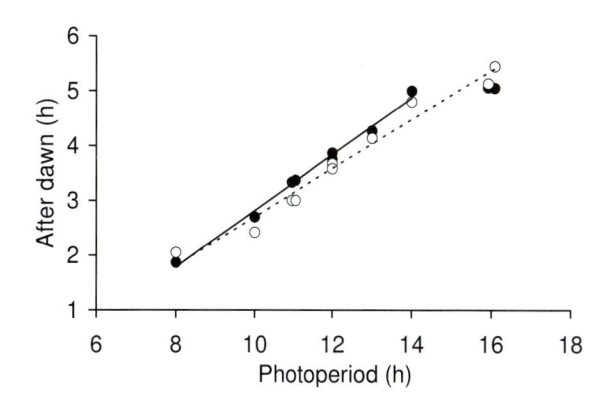

Figure 3.19 *Mean oviposition time (●) and time when 0.50 of eggs have been laid (○) for broiler breeders given photoperiods of 8 to 16 h (55).*

of feed. Body weight at 20 weeks has been reported to decrease by 14 g for each 1-h of daylength, which is most likely a consequence of higher energy expenditure during light than dark (54). However, rearing on longer photoperiods results in an increase of 60 g in body weight at 50% lay because the birds are older (53). As a consequence, feed required to reach this level of egg production increases by 630 g for each extra 1 h of daylength.

Egg production to a typical broiler breeder depletion age of 60 weeks is closely linked to age at sexual maturity, decreasing by 4 to 7 eggs for each 10-d delay in sexual maturity (52, 53). It is also influenced by the incidence of adult photo-refractoriness because birds spontaneously stop egg-laying after prolonged exposure to long days. It is not surprising, therefore, that maximum egg numbers are produced when broiler breeders are maintained on 10-h daylengths (Figure 3.18).

Egg weight
Mean egg weight over the laying cycle increases by 0.4 g per 1 h of photoperiod (53). This rate is about double that reported for *ad libitum*-fed egg-type hens (3). However, the response can be partitioned, as follows, into a direct photoperiodic effect and indirect influences of age and body weight at sexual maturity.

$$y = 52.0 + 2.260BW + 0.048A + 0.109p$$
$$(r^2 = 0.969, P<0.001, SD = 0.21)$$

where y = mean egg weight to 60 weeks adjusted for growth rate during the rearing phase, BW = body weight at 50% lay (kg), A = age at 50% lay (d) and p = photoperiod (h).

The 0.11 g per h rate of increase, which may be regarded as the true photoperiodic effect, agrees reasonably well with the regression for full-fed egg-type hybrids (3) and suggests that the direct photoperiodic influence on egg weight is independent of genotype. A stepwise regression of the published data showed that age at 50% lay was responsible for 0.73 ($P<0.001$), body weight at 50% lay for 0.22 ($P<0.001$) and photoperiod for only 0.02 of the effect ($P=0.056$).

Feed conversion efficiency
The net result of the photoperiodic effects on sexual maturity, egg numbers and egg weight is that hens maintained on 13 or 14-h daylengths use less feed per gram of egg than those on other daylengths (53).

Time of lay
The effect of daylengths between 8 and 14 h on mean time of egg-laying in broiler breeders is very similar to that for egg-type hybrids, being delayed by about 30 min for each 1 h of photoperiod (Figure 3.19, 55). However, the actual mean oviposition time (MOT) occurs 2 h later than in egg-type hybrids for a given photoperiod, and is described by the equation:

$$MOT = -2.34 + 0.52p$$
$$(r^2 = 0.991, P<0.001, SD = 0.10)$$

where p = photoperiod (h).

In the same way that MOT is earlier in modern egg-type hybrids as a consequence of indirect selection for more rapid egg formation, so the limited selection for egg numbers in broiler breeders means that they take longer to form an egg and so lay later in the day. However, the MOT agrees well with historical data for white-egg hybrids (56, 57). Whereas modern egg-type hybrids now lay a proportion of their eggs before dawn, on daylengths up to 16 h, pre-dawn egg-laying by broiler breeders is negligible when a daylength of at least 13 h is provided (55).

Shell quality
Shell weight and shell thickness index are both negatively correlated with photoperiod, with shell weight decreasing by 30 mg and thickness index by 0.57 mg/cm² for each 1-h increase in daylength (58). As suggested for egg-type hens (page 28), it is likely that the superior shell quality on shorter photoperiods is a result of prolonged activity during the scotoperiod of parathyroid hormone and calcitonin on calcium resorption from the skeleton, with the opposite scenario resulting in inferior shells when birds are exposed to longer daylengths but shorter nights. These effects on shell quality may have consequences for hatchability (59).

• Broiler breeders: effects of changing photoperiods •

Although growth exerts its own effect on, but does not interact with, the reproductive response to constant photoperiods, it does interact profoundly with a broiler breeder's response to an increase in daylength. One reason is that a bird cannot respond to an increase in light until it has fully dissipated juvenile photo-refractoriness, and the rate at which this is achieved is dependent on the degree to which body weight is controlled. Another is that the degree of feed restriction necessary to control growth for satisfactory subsequent egg production is such that its influence is ten-fold that of the effect of the age at which a broiler breeder is photostimulated. Sexual maturation in dwarf broiler breeders is totally suppressed if the feed allocation is reduced to a level that does not allow growth beyond 1 kg (60). Rate of sexual development also depends on the body weight growth curve, with birds grown to achieve typical broiler breeder body weights (almost linear) maturing 4 d earlier than birds permitted faster initial growth and slower growth from about 10 weeks to reach the same 20-week body weight (61).

Broiler breeders are allocated fixed quantities of feed, and so the only aim of a prepubertal increment in photoperiod is to control sexual maturity. As for egg-type hybrids, there is a critical photoperiod that needs to be exceeded to stimulate GnRH and gonadotrophin release and a saturation photoperiod at which gonado-trophin release is maximised. Figure 3.20 shows that the response of normal broiler breeders is similar to that for dwarf stock, and that the critical daylength for birds reared on 8 h is between 10.5 and 12.75 h and the saturation daylength is between 12.75 and 15.25 h.

An analysis of data from several experiments conducted at the University of KwaZulu-Natal shows that the first pullet in a flock grown according to a typical primary breeding company's recommendations does not become photosensitive until about 10 weeks of age and that the flock does not respond as a whole until it is about 18 weeks old (Figure 3.21). Before photorefractoriness has been dissipated, the bird will respond as if it had been maintained on the final photoperiod from hatch, and so pullets

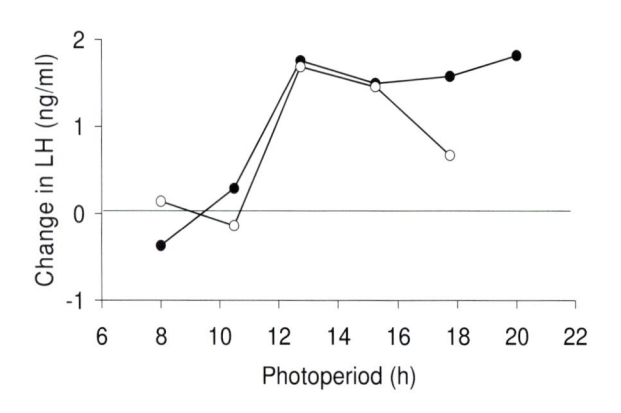

Figure 3.20 *Change in plasma LH 6 d after a transfer from 8 h to various photoperiods for normal size (○) and dwarf broiler breeders (●) (14 and unpublished).*

transferred from 8 to 16 h when they are younger than 10 weeks will be about 3 weeks later maturing than constant 8-h controls (see Figure 3.17), whereas birds photostimulated at 18 weeks will have their maturity advanced by about 5 weeks (52). Photostimulation between 10 and 18 weeks results in a bimodal distribution of ages at first egg, with some birds having their maturity advanced and others having it delayed. This period from 10 to 18 weeks during which photosensitivity is acquired in broiler breeders is double the time taken for full-fed egg-type pullets, which start to be photosensitive at about 5 weeks and are fully responsive by 9 weeks. However, both types of stock are about 0.2 of mature body weight when they start to become photosensitive and 0.4 when they are completely responsive. This indicates that photosensitivity occurs according to physiological not chronological age, and so providing broiler breeders with more feed will accelerate the attainment of photosensitivity and reducing feed will delay it. A regression analysis of the maturity data for the various body-weight groups in Figure 3.21 shows that, once a group has become fully photosensitive, there is no significant difference in the rate at which

Table 3.2 *Effect of photoperiod during the rearing period on sexual maturity and egg production in broiler breeders transferred to 16 h at 20 weeks (62).*

Trait	Photoperiod (h)		
	6	8	10
Age at 50% lay (d)	197	194	198
Eggs to 60 weeks	164	165	167
Egg weight (g)	68.1	67.7	68.2

maturity is retarded by delaying transfer to a stimulatory daylength. There is, however, a difference in the elevation of the line; heavier birds have earlier maturity. The common slope for all body weights is 0.44 d delay for each 1-d later photostimulation, which compares well with the 0.49-d rate of change for brown-egg pullets transferred from 8 to 16 h (Figure 3.4).

The effect of rearing daylength is negligible when broiler breeders are transferred from 6, 8 or 10 h to a stimulatory photoperiod at 20 weeks. Age at 50% lay differs by only 3 or 4 d and subsequent rates of egg production are similar (Table 3.2, 62). The effects on sexual maturation are similar to those predicted for

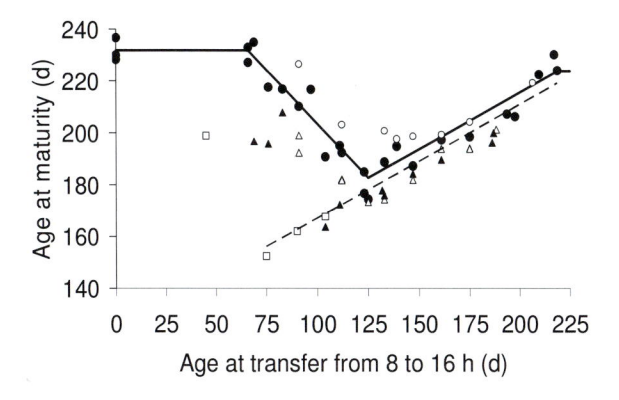

Figure 3.21 *Mean age at sexual maturity for broiler breeders grown to achieve a mean 20-week body weight of 1920 (○), 2130 (●), 2610 (△), 2820 (▲) or 3260 (□) g and transferred from 8 to 16-h photoperiods at various ages. The solid line represents the mean response for pullets grown to achieve a typical 20-week body weight of 2130 g. The dotted line represents the common slope for all body-weight groups determined by a linear regression with differences from the 2130-g group removed by least squares analysis (unpublished data from the University of KwaZulu-Natal).*

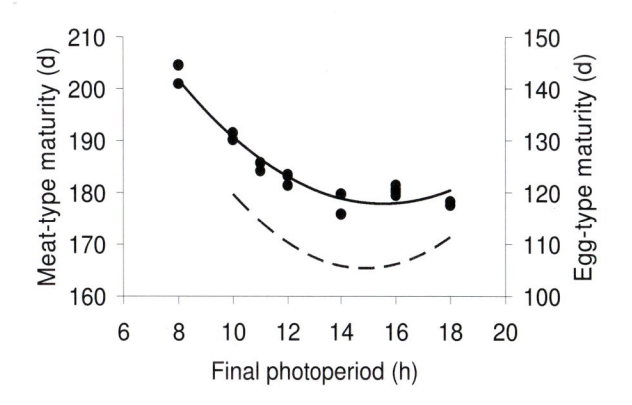

Figure 3.22 *Effect of final photoperiod on age at sexual maturity for broiler breeders grown to achieve about 2100 g body weight at 20 weeks and transferred from 8-h photoperiods at 140 d, with the difference between trials removed by least squares (52, 63). The broken line represents the regression for egg-type modern hybrids photostimulated at 70 d, a similar physiological age to 140-d broiler breeders (unpublished data from the University of KwaZulu-Natal).*

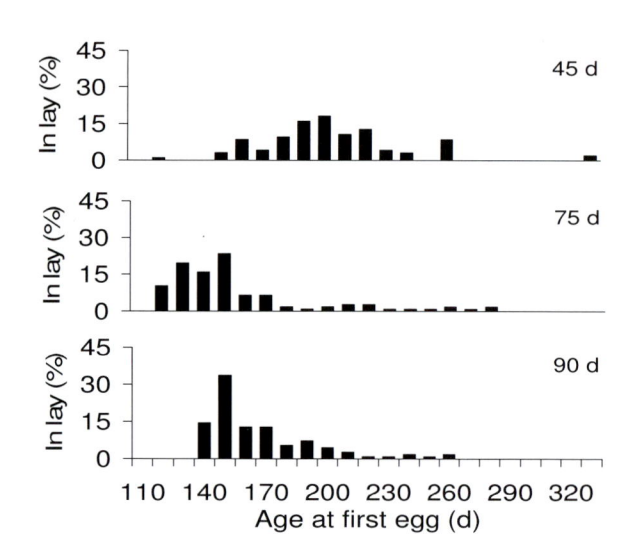

Figure 3.23 *Proportion of broiler breeders in 10-d classes laying their first egg in response to a transfer, at 2100-g mean body weight, from 8 to 16-h photoperiods at 45, 75 or 90 d (unpublished data from the University of KwaZulu-Natal).*

ad libitum-fed egg-type hybrids photostimulated at 10 weeks (the same physiological age).

There is a dearth of information on the effect of transferring broiler breeders to different photoperiods, and the only data available are for birds reared on 8-h photoperiods and transferred to various final daylengths at 20 weeks (52, 63). Consequently, in contrast to egg-type pullets, it has not been possible to create a full model to predict the response function for meat-type stock. Nevertheless, the data in Figure 3.22 show that the effect of final photoperiod on the photosexual response is similar for both types of chicken, with transfers to 14 or 16 h inducing the earliest maturity. It seems reasonable, therefore, to use the *b* values from the egg-type pullet model (page 26) for predicting responses for broiler breeders.

Figures 3.21 and 3.23 show that accelerating growth is one way of advancing sexual maturity, but the need for sufficient time to dissipate juvenile photorefractoriness will dictate how early a pullet can be photostimulated, and this is likely to be about 18 weeks if bimodal responses are to be avoided. Although acceleration of growth has resulted in advances in mean age at first egg, if birds are changed to a stimulatory photoperiod as early as 45d, sexual development will not be initiated in all birds in the group at this age. Those that are either too physically immature to respond or have not had their photorefractoriness dissipated will mature as if maintained on the final photoperiod, resulting in a bimodal distribution with some birds not laying their first egg until they are more than 40 weeks of age (Figure 3.23).

There are insufficient data to construct a model to predict age at sexual maturity in broiler breeders that is as robust as that for egg-type pullets, but a multiple regression using 62 data sets for age and body weight at photostimulation, body weight at 20 weeks and final photoperiod from 10 trials conducted at the University of KwaZulu-Natal between 2001 and 2005 for birds reared on an 8-h photoperiod gives a good estimate. Body weight at photostimulation and at 20 weeks are included in the model because the shape of the growth curve as well as body weight *per se* influences the timing of sexual maturation (61). Age at first egg data were converted to an age at 50% lay (ASM) equivalent by multiplying by 1.0295. The prediction equation for ASM is:

$$y = 453.5 - 1.293PA + 2.223^{-9}PA^5 - 9.502^{-12}PA^6$$
$$- 0.0414PB + 1.244^{-5}PB^2 - 0.0365BW - 1.6639p$$
$$(r^2 = 0.944, P<0.0001, SD = 5.02)$$

where y = mean age at 50% lay (d), PA = age at photostimulation (d), PB = body weight at photostimulation (g), BW = body weight at 20 weeks (g) and p = final photoperiod (h).

Using data generated by the model, Figure 3.24 shows the extent to which a 10% increase or decrease in a typical 20-week body weight of 2160 g alters the youngest age at which birds can be transferred from 8 to 16-h daylengths to maximise the advance in maturity. The lowest points on each curve indicate that increasing 20-week body weight by 100 g lowers the youngest age for successful photostimulation by about 2 d and earliest attainable ASM by about 3.5 d.

It is good practice to delay photostimulation of under-weight pullets until they have reached 2.0 kg, otherwise they will have both retarded maturity and sub-optimal egg production (64).

There are few reports of the influence of final photoperiod on egg production but limited data suggest that there is a direct photoperiodic effect

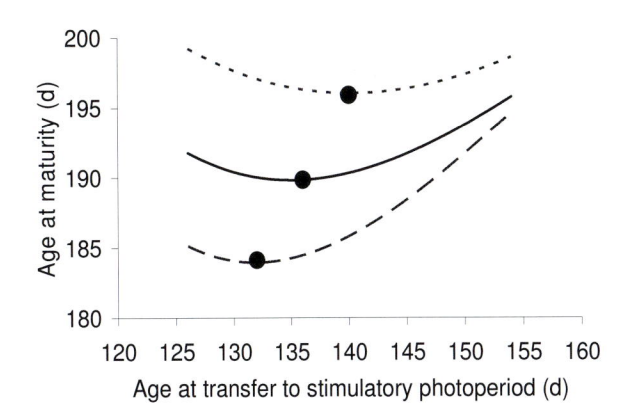

Figure 3.24 *Effect of age at transfer from 8 to 16-h photoperiod on age at sexual maturity for broiler breeders grown to achieve a 20-week body weight of 1940 g (0.9 of conventional body weight, dotted line), 2160 g (conventional, solid line) or 2380 g (1.1 conventional, broken line), and the youngest photostimulation age to achieve the earliest maturity (●).*

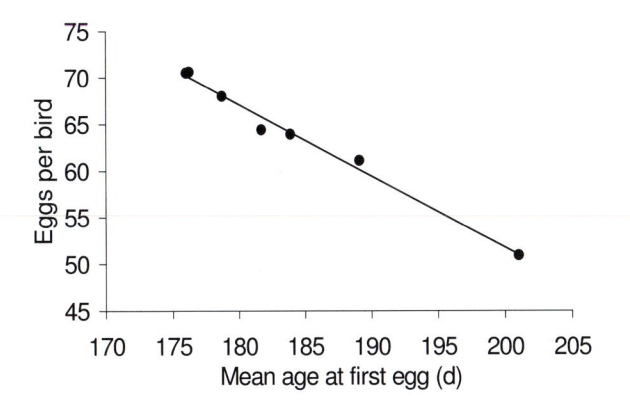

Figure 3.25 *Effect of age at first egg on egg numbers to 39 weeks of age (63).*

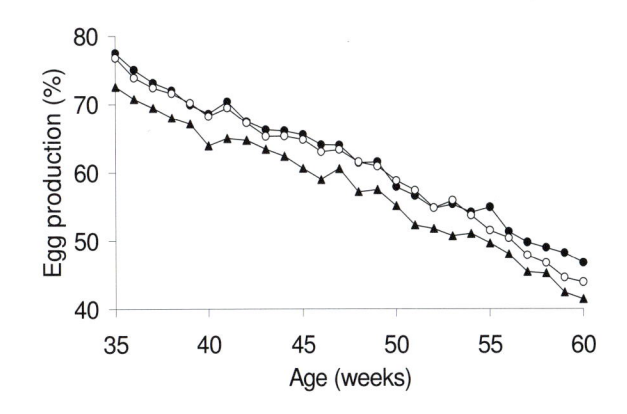

Figure 3.26 *Egg production for broiler breeders maintained on 11 (●) or 16-h (▲) photoperiods from 20 weeks, or given further increments to 16 h between 35 and 54 week (○) (unpublished data from the University of KwaZulu-Natal).*

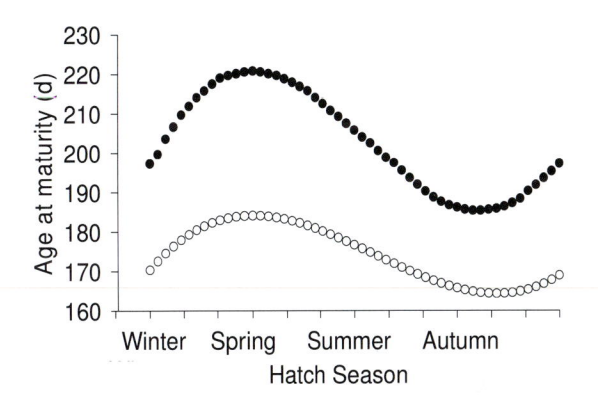

Figure 3.27 *Effect of hatch-date on age at sexual maturity for broiler breeders located at 42°N in Spain (●, 66) and at 20-25°N in Mexico (○, 67).*

that is independent of the indirect effect of age at sexual maturity (52). Nevertheless, the data in Figure 3.25 show that egg production, at least during the first half of the laying cycle, is strongly influenced by age at first egg, decreasing by about 7 eggs for each 10-d delay in maturity (63).

Hens transferred to long (e.g. 16 h) daylengths have inferior post-peak rates of lay compared with hens transferred to moderate (11 or 12 h) photoperiods (Figure 3.26, unpublished data from the University of Kwa-Zulu-Natal,), and this results in the production of fewer eggs to a typical depletion age of 60 weeks, despite having had the advantage of earlier sexual maturation.

Whilst this may be partly explained by the higher energy expenditure of birds on longer daylengths, some hens on 16-h photoperiods go temporarily out of lay, whilst others cease egg laying prematurely due, probably, to the onset of adult photorefractoriness (52).

Broiler breeders do not respond positively to increases in photoperiod given during lay. There is no response if they are already on long daylengths (65), and increments given to hens held from 20 to 35 weeks on an 11-h day can actually depress rate of lay (Figure 3.26).

Natural light

Broiler breeders in non-controlled environment housing and hatched at different times of the year mature differently despite being managed according to a breeding company's recommendations. Birds hatched in the spring mature the latest and those hatched in the autumn mature the earliest. The variation depends not only on the light-tightness of the house but also on latitude, with flocks located at higher latitudes showing larger variation because of the greater contrast between the shortest and longest day (Figure 3.27).

• Broilers: effects of constant and changing photoperiods •

Traditionally, broilers have been given long daylengths or continuous illumination to maximise feed intake and growth. In the early days of the broiler industry, body weight was 5 to 10% lower for birds exposed to 8 or 12-h photoperiods compared with birds given permanent light, but those given only 6-h days learned to eat in the dark and grew at a faster rate than birds given 14-h photoperiods (1). In the extreme, broilers transferred to continuous darkness at 7 d of age were reported to have a similar body weight at 63 d to birds given continuous illumination (68). However, data in Table 3.3 derived from published (69) and unpublished research findings (S. H. Gordon, personal communication) show that whereas body weight gain in modern broilers increased by about 10 g per 1 h of photoperiod up to 21 d, it decreased by about 6 g/h between 22 and 49 d for birds given ≥12 h, though birds on 8-h daylengths continued to have the slowest growth. As a consequence, body weight at 49 d, with the exception of 8-h birds, was similar for all daylengths. This is because, although feed intake during the first 21 d increased linearly by about 15 g/h of photoperiod, there were no significant differences in total feed intake to 49 d for photoperiods of 12 h or longer. Broilers held on the 8-h day ate less feed and gained less weight but, at 49 d, there were no significant differences in feed conversion efficiency between any of the photoperiods. In contrast, photoperiod strongly influenced bird health, with longer daylengths being associated with a higher incidence of Sudden Death Syndrome, leg disorders and overall mortality.

When 1970s broilers were given moderate daylengths (12 to 14 h) at an intensity of 350 lux, feeding activity was highest during the first hour of the photoperiod, but when the light intensity was increased progressively from 0 to 350 lux over the first 2 h of the photoperiod and decreased during the last 2 h, to simulate a 2-h dawn and dusk, feeding activity peaked during the last two hours of illumination, feed intake increased by 5% and body weight was 2% higher at 63 d (71). These findings suggest that the broiler can predict the end of the day more effectively if there is dusk, and that this facilitates an increase in crepuscular feeding activity to ensure a full crop at the beginning of the night and an overall increase in feed intake to match that of continuously illuminated controls. The reduced energy expenditure of the birds on the dawn-dusk regime then enabled them to use the feed more efficiently and achieve the heavier body weight. However, just as modern broilers respond to daylength differently from early stock, so modern birds might react differently to a dawn-dusk programme from their ancestors of 30 years ago.

Whereas the provision of 6-h photoperiods reduces body weight at, and cumulative feed intake to, 21 d compared with birds grown on 23-h photoperiods, a transfer from short to long days results in subsequent compensatory growth so that body weights for the two groups are similar by 42 d (Table 3.4). However, the initial 3 weeks of 6-h days reduces overall mortality and the incidence of leg disorders through to 42 d, and produces comparable liveability figures to birds maintained on short days throughout but about half of those for birds given 23-h daylengths (Table 3.5, 72, 73). Skeletal development is related more to age than to body weight (74), and so retarding initial growth by the provision of short photoperiods for the first one to three weeks results in smaller loads on the leg joints. The skeleton is then better able to carry the rapid growth that occurs when the birds are

Table 3.3 *Body weight, feed intake and feed conversion efficiency for male and as-hatched broilers fed ad libitum and given various constant photoperiods between 8 and 23 h (70). Differences between the male and as-hatched birds were removed by least squares analysis.*

	Photoperiod (h)					
	8	12	14	16	20	23
0 to 21 d						
Feed intake (g)	840	905	945	1010	1050	1070
Body weight (g)	660	695	720	760	790	815
Feed conversion (g/g)	1.36	1.39	1.40	1.41	1.41	1.40
22 to 49 d						
Feed intake (g)	3865	4045	3840	3930	3850	3905
Body weight gain (g)	1765	1835	1775	1780	1760	1750
Feed conversion (g/g)	2.19	2.21	2.17	2.21	2.19	2.23
0 to 49 d						
Feed intake (g)	4705	4950	4785	4940	4900	4980
Body weight gain (g)	2425	2530	2495	2540	2550	2565
Feed conversion (g/g)	1.99	2.02	1.99	2.01	2.01	2.01
Total mortality (%)	6.7	8.5	11.8	11.4	16.3	17.3
Sudden death (%)	1.6	2.6	*	4.0	6.2	*
Leg disorders (%)	0.4	0.3	*	0.8	1.1	*

Table 3.4 *Body weight, feed intake, feed conversion efficiency and mortality for as-hatched broilers given an initial 3 d of continuous illumination and then transferred permanently to 23 h or given a 6-h photoperiod between 4 and 21 d, with a transfer to 23 h at 21 d. Data are from 3 trials where the transfer was made abruptly and 5 trials where it was made gradually. Differences among trials removed by least squares analysis (70).*

	Constant 23-h	0-21 d, 6 h 22-49 d, 23 h	Paired *t* test *P* value
0 to 21 d			
Feed intake (g)	985	855	<0.001
Body weight (g)	705	635	0.002
Feed conversion (g/g)	1.41	1.34	<0.001
0 to 42 d			
Feed intake (g)	3840	3745	0.036
Body weight (g)	2105	2095	0.662
Feed conversion (g/g)	1.83	1.80	0.005
Total mortality (%)	10.8	6.3	0.001
Sudden death syndrome (%)	4.8	3.3	0.029
Leg disorders (%)	8.6	6.0	0.027

Table 3.5 *The effect of an abrupt transfer from 6 to 23 h at 21 d or a gradual transfer to 23 h between 14 and 35 d on the incidence of Sudden Death Syndrome, leg disorders and total mortality in male (M) and female (F) broilers to 42 d compared with constant 23-h controls (72).*

	Constant 23 h M	Constant 23 h F	Abrupt transfer from 6 to 23 h at 21 d M	Abrupt transfer from 6 to 23 h at 21 d F	Gradual transfer from 6 to 23 h from 14 d M	Gradual transfer from 6 to 23 h from 14 d F
Sudden Death Syndrome (%)	2.4	0.7	1.4	0.5	1.4	0.1
Leg disorders (%)	4.7	1.8	2.5	0.9	2.2	1.1
Other mortality (%)	1.0	1.8	0.9	0.9	1.1	0.7
Total (%)	8.1	4.3	4.8	2.3	4.7	1.9

transferred to long photoperiods. Although the benefits are greater for males than for females, they are proportionally the same for both sexes (Table 3.5). Improvements in the skeletal integrity of meat-birds initially exposed to shorter photoperiods are really responses to longer scotoperiods and, as for the benefits to shell quality in laying hens, prolonged nocturnal melatonin secretion results in an extended release of calcitonin and parathyroid hormone, and an enhancement of calcium mobilisation and advantageous modifications of osteoclast and osteoblast activity (24, 25, 26). However, ultra short daylengths may adversely affect skeletal health because long periods of inactivity could compromise the blood supply to the leg-bone growth plates (75). Some of the compensatory body weight gain that follows a change to long daylengths, especially ≥ 5 weeks (76), may be in response to an increase in plasma testosterone, as indicated by male birds having larger and brighter combs than constant-photoperiod or naturally lit controls at 42-49 d (72).

• Growing turkeys: effects of constant photoperiods •

Although feed intake and body weight gain in turkeys during the first few weeks tend to be related to daylength, especially when it exceeds 12 h, by 6 weeks there is little effect of daylength on either trait. This is because turkeys on short days, like broilers, learn to eat in the dark (Table 3.6). Birds on 8-h photoperiods progressively increase their feed intake to match or exceed that of birds given longer daylengths, and body weights among the different photoperiods equalise (Table 3.7). However, between 12 and 20 weeks, male turkeys maintained on stimulatory daylengths (≥ 12 h) start to develop sexually and, in response to increasing concentrations of plasma testosterone, convert feed more efficiently and have larger body weight gains during this phase (Table 3.7, Figure 3.28). These data indicate that growth during this period of sexual maturation is proportional to the stimulatoriness of the photoperiod and not simply its duration.

Table 3.6 *Rate of feed intake during the light and darkness at 14 weeks for male turkeys given constant 8, 12, 16 or 23-h photoperiods (unpublished data from the University of Bristol).*

Photoperiod (h)	Hourly intake (g) Light	Dark	Daily mean
8	23.4	21.2	21.9
12	29.0	16.4	22.7
16	31.6	0.2	21.1
23	23.3	0.0	22.1

The increased circulating testosterone induces faster testicular growth, and unpublished data from the University of Bristol show that each 10-g increase in testicular weight is associated with a 650-g increase in body weight gain between 15 and 20 weeks.

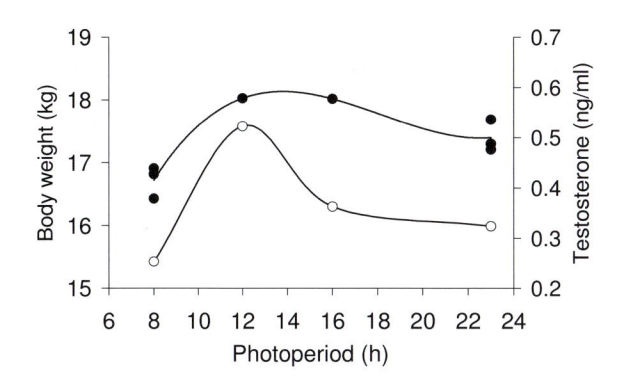

Chapter 3 Photoperiod: Conventional programmes

Figure 3.28 *Plasma testosterone concentration at 15 weeks (○) and body weight at 20 weeks (●) for male turkeys given constant 8, 12, 16 or 23-h photoperiods (77, 78, unpublished data from the University of Bristol).*

Table 3.7 *Body weight, feed intake and feed conversion at 108d, and weight gain, feed intake and feed conversion between 108 and 136 d for male turkeys given constant 8, 12, 16 or 23-h photoperiods (77).*

Trait	Photoperiod (h)			
	8	12	16	23
0-108 d				
Feed intake (kg)	32.4	32.7	31.7	31.3
Body weight (kg)	12.9	13.4	12.7	13.0
Feed per kg body weight gain (kg)	2.53	2.44	2.49	2.42
108 – 136 d				
Feed intake (kg)	15.8	17.4	17.2	15.4
Weight gain (kg)	3.56	4.63	5.28	4.68
Feed per kg body weight (kg)	4.44	3.77	3.23	3.28

Feed conversion efficiency to 20 weeks improves curvilinearly with photoperiod in males, but there is no discernible improvement beyond 16 h, (Figure 3.29).

Longer photoperiods have an adverse effect on the number of deaths and culls in intact (not beak trimmed, de-toed or de-snooded) male turkeys, even when the light intensity is reduced to only 1 lux to control cannibalistic behaviour (77). However, when males are beak trimmed, toe-clipped and de-snooded, daylength does not seem to affect liveability (Table 3.8, 78).

Further evidence of the photosexual influence on feed intake, growth and mortality in male

Table 3.8 *Deaths and culls to 20 weeks for intact male turkeys given 8, 12, 16 or 23-h photoperiods at an illuminance of 1 lux (77), and, in a separate trial, to 22 weeks for beak-trimmed, de-toed and de-snooded males given 8 or 23-h photoperiods (78).*

Trait	Photoperiod (h)			
	8	12	16	23
Culls (%)	0	1	1	5
Intact birds				
Deaths (%)	4	8	9	7
Total losses (%)	4	9	10	12
Beak-trimmed, de-toed & de-snooded				
Total losses (%)	17.2	*	*	16.7

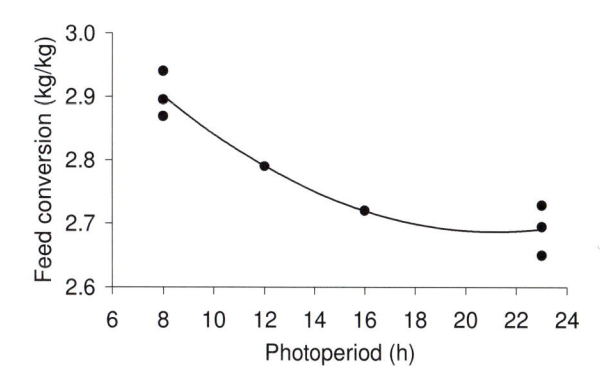

Figure 3.29 *Feed conversion ratio (kg feed/kg body weight) for male turkeys grown to >20 weeks, differences between trials removed by least squares analysis (78, 79).*

turkeys is provided by the heavier body weight but higher mortality at 17 weeks of turkey toms given a step-up programme from 8 to 23 h compared with constant 23-h controls (79).

Research findings suggest that the effects of constant daylength on growth are much less important for females than for males.

Seasonal effects

Simulated seasonally changing photoperiods (at 36°N and at an illuminance of 22 lux) do influence performance, and the effect is independent of the prevailing season. Simulated autumn-hatched birds had heavier body weights and better feed conversion efficiency than simulated spring-hatched turkeys through to 16 weeks, whether the birds were actually hatched in spring or in autumn. However, trials conducted in the spring to summer period produced heavier body weights and more efficient feed conversion than birds grown during the autumn and winter; presumably as a consequence of the different temperatures in these seasons (80).

Various high-illuminance (20 lux) step-up and low-illuminance (2.5 lux) step-down lighting regimes have been used to improve leg integrity in both males and females (81). Generally, the incidence of leg problems has been reduced by step-up lighting; possibly as a result of increased exercise. However, the effects seem to be dependent upon light intensity, because no benefits to leg health were observed when step-up lighting was given at 2.5 lux (82). Body weight and feed conversion efficiency responses have been variable but, where there have been significant influences, they appear to have been consequences of the effect that the current photoperiod had on feed intake. For example, when 30-min weekly increases or decreases were given from 4 to 16 weeks between 10 and 16-h photoperiods, hens on the step-down regime initially had a larger feed intake and faster growth than step-up birds, but at the end, when their daylength was shorter, feed intake and body weight gain were inferior; intakes and body-weight gains were similar in the middle period when the daylengths were similar (83).

Female turkeys mature later and are usually killed at younger ages than males and, as a consequence, do not show the steroid-induced accelerated growth that occurs in males exposed to a stimulatory lighting regime. However, oestrogen does influence growth if hens are kept to an age that permits a photosexual response.

• Breeding turkeys: effects during growing and laying periods •

Turkeys exhibit juvenile photorefractoriness, and therefore need to be given a period of short daylengths at the end of rearing to enable them to become responsive to a stimulatory photoperiod and have satisfactory egg production. There can be an interaction between the daylength and the length of time that short days need to be given to dissipate photorefractoriness. For example, 3 weeks of 4-h photoperiods are reported to equate to 4 weeks of 6-h or 5 weeks of 8-h day (84). In commercial practice, it is recommended to give the modern turkey 10 to 11 weeks of 6- or 7-h days starting at 18 weeks.

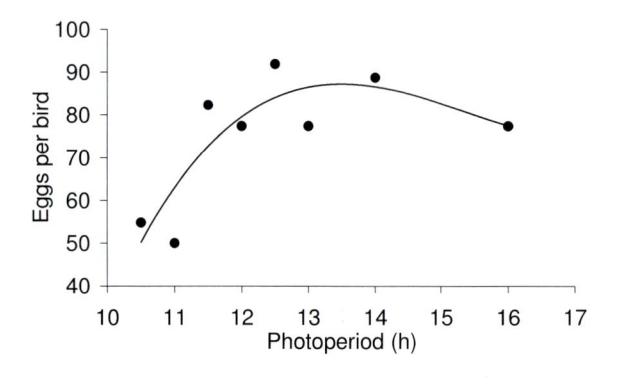

Figure 3.30 *Cumulative egg numbers to 54 weeks for turkey hens transferred from 8 h to various photoperiods between 10.5 and 16 h at 30 weeks (85).*

The critical daylength required to successfully induce rapid gonadal development in photosensitive females is reported to vary with season, being < 10.5 h for 30-week-old birds photostimulated in winter but 11 h in summer (85). Interestingly, these findings agree with observations for turkeys exposed to natural lighting (35°N) where first eggs were laid when the daylength (sunrise to sunset) was about 11.5 h (86). The minimum photoperiod to optimise egg production also varies with season, and is about 30 min longer than the critical daylength: 11 h for winter stimulated birds and 11.5 h for those stimulated in summer.

As with all seasonal breeding birds, prolonged exposure to long daylengths eventually results in gonadal regression. In both winter and autumn photostimulated flocks, the critical daylength for the onset of adult photorefractoriness is about 12 to 12.5 h (86, 87).

Figure 3.30 shows the effect of transferring photosensitive hens from 8 h to various photoperiods on egg production in a 24-week cycle (85). The data show that egg production is maximised between 11.5 and 14 h but may be depressed if daylength is extended further. Although turkeys transferred to longer photoperiods have earlier and higher peak rates

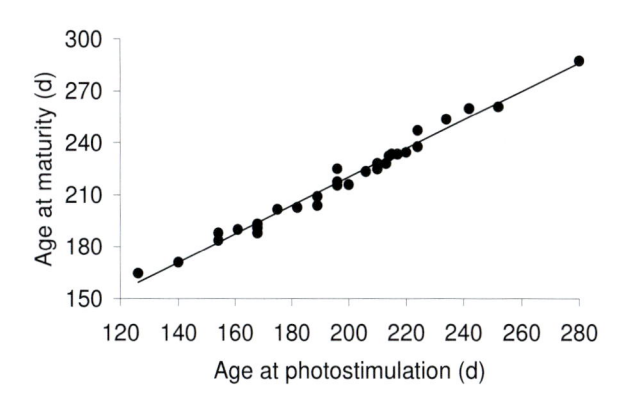

Figure 3.31 *Mean age at sexual maturity for turkey females transferred from a short day to a stimulatory photoperiod (usually 14 h) at various ages between 18 and 40 weeks (89).*

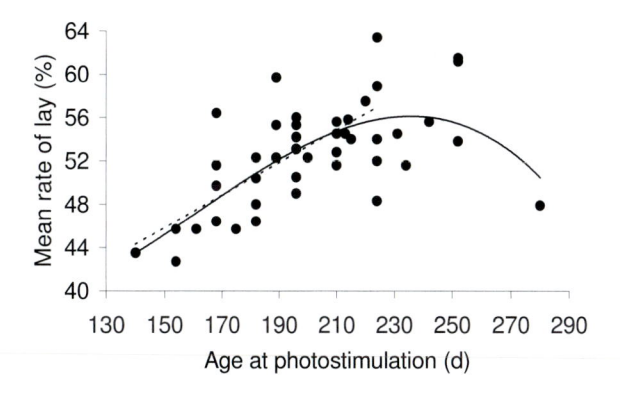

Figure 3.32 *Mean rate of lay for a fixed period from photostimulation for turkey females transferred to a stimulatory photoperiod (usually 14 h) at various ages between 18 and 40 weeks (89). The broken line represents the linear regression for photostimulation ages between 20 and 32 weeks.*

of lay, poorer persistency eventually results in the production of fewer total eggs (85). This is similar to the response of broiler breeders to daylength during the laying period (page 37). It also fits with findings that the critical daylength for the onset of adult photorefractoriness in the turkey is about 12 h, and the suggestion that egg production is maximised by a daylength that is long enough to provide peak photoperiodic drive but short enough to delay the onset of gonadal regression (87). There appears to be no advantage in giving turkeys a series of increments over an abrupt increase to the final photoperiod (88).

The age at which turkeys are transferred to the final photoperiod, as for egg-type (Figure 3.4, page 25) and broiler breeder (Figure 3.21, page 35) pullets, strongly influences the age at first egg and rate of lay (Figures 3.31, 3.32; 89). A comparison of the response rate of turkeys with that of egg-type hybrids and broiler breeders following a transfer from 8 to 14-h days shows the turkey to be more responsive. For each 1-d delay in photostimulating, sexual maturity in the turkey is delayed by 0.8 d, whilst in the domestic fowl it is delayed by 0.5 d.

However, the rate of change in turkey maturity for a given delay in photostimulation is similar to responses of 0.82 d for partridges (90) and 0.88 d for quail (91).

As for egg-type hybrids (92), there is a negative correlation of egg numbers to a fixed age with age at photostimulation but, initially, a positive link for the number of eggs laid in a given period from sexual maturity (93). However, the improvement in egg production for a given production period is curvilinear, with maximum egg numbers being produced when turkeys are transferred from short to long days at about 32 weeks (89). For practical purposes, the effect of age at photostimulation on egg production can be treated as being linear between 20 and 32 weeks, with mean rate of lay increasing by about 2% for each 10-d later maturity or, because age at sexual maturity is linearly related to age at photostimulation, by about 1.5% for each 10-d delay in photostimulation (Figure 3.32).

Mean egg weight increases linearly by 1.3 g for each 10-d later photostimulation, by approximately 1.6 g for each 10-d delay in sexual maturity or by 1.1 g for each 100 g increase in body weight at sexual maturity. Although these rates of change are double those for domestic fowl in absolute terms, they are similar when related to body weight at sexual maturity (90).

• Waterfowl •

There are few reports of research findings for lighting waterfowl, even though much of the original research into the avian photoperiodic response was conducted using Mallard drakes. Some waterfowl are seasonal breeders and some are not, and this will have a large influence on their respective photoperiodic responses. Accordingly, the various species and sub-species will be discussed separately.

Muscovy (Barbarie)

Changes in plasma LH concentration over a 12-week period following a transfer from 6 to 7-h photoperiods at 15 weeks or by providing a 1-h light pulse to create a 9, 11 or 13-h subjective day (Figure 3.33) indicate that the critical daylength for a photosexual response in the Muscovy drake lies somewhere between 11 and 13 h (94). This range agrees with the findings for females (95) and is similar to that in broiler breeder pullets (Figure 3.20).

The Muscovy female does not appear to suffer juvenile photorefractoriness because *ad libitum*-fed ducks maintained on 16-h daylengths for 20 weeks then stepped down to 13 h and back to 16 h reached 10% lay 6 weeks earlier than birds maintained on 8 h and stepped up to 14 h at 20 weeks, and ducks reared on 13 h to 20 weeks and then increased to 15 h reached 10% lay 5 weeks before constant 8-h controls (95). These findings are not consistent with the responses of photorefractory birds.

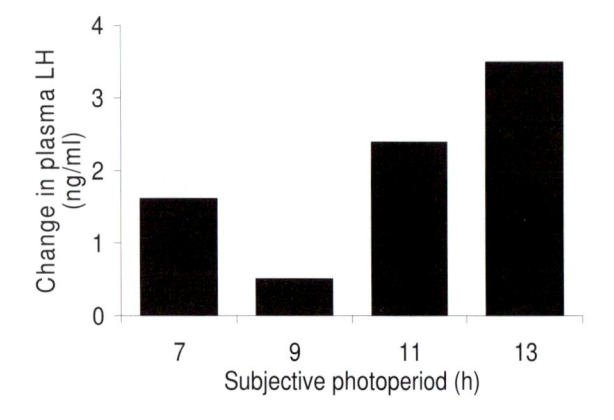

Figure 3.33 *Change in plasma LH between 15 and 27 weeks of age for Muscovy drakes transferred from 6 to 7, 9, 11 or 13-h subjective photoperiods (94).*

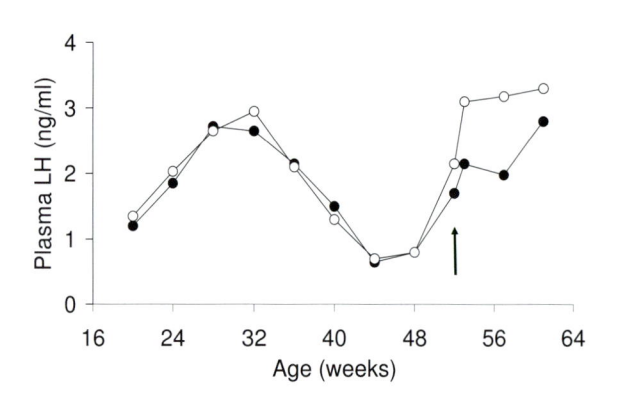

Figure 3.34 *Plasma LH concentration for Muscovy drakes transferred from 6 to 16-h photoperiods at 20 weeks and maintained on 16 h (●) or given a further increase to 20 h at 52 weeks (○), with the second photostimulation indicated by the arrow (96).*

Prepubertal lighting history also appears to be far less important for Muscovy than for domestic fowl or turkeys. Age at 50% lay for birds stepped down from 24 h to a constant 14 h from 26 weeks was similar to that for birds reduced from 24 to 8.5 h at 4 weeks and then given weekly increases of 30 min to also reach 14 h at 26 weeks (95). Although these data suggest that Muscovy ducks are only mildly photoperiodic, possibly because in their tropical origins other environmental cues exerted more control over reproductive activity than light, they still do respond to an increment in daylength, especially if it follows a period of 6 or 8-h days. The rate of change in age at 50% lay for a given delay in photostimulation is similar to that in egg-type hybrids and broiler breeders, with a 1-week delay in maturity resulting from a 2-week later transfer from 8 to 14 h (95).

In contrast, Muscovy drakes, which following a transfer from 6 to 16-h daylengths at 20 weeks have been observed to undergo testicular regression and enter a moult within 24 weeks of exposure to long-days, seem to exhibit some form of adult photorefractoriness. However, subsequent testicular recrudescence and increases in plasma LH and testosterone concentrations without any change in photoperiod, and a photoinduced rise in plasma LH when the 16-h photoperiod was further extended to 20 h do not support this conclusion

(Figure 3.34, 96). Similar phenomena are seen in other duck species (98, 99) and in turkey hens (97); the condition in turkeys has been termed *relative* photorefractoriness. However, spontaneous recrudescence whilst still on long-days does not suggest that gonadal regression occurred because unfavourable conditions for rearing offspring were anticipated (true adult photorefractoriness), so the condition might be more accurately called a 'mid-season pause'.

Geese

Emden, Toulouse and White Roman geese are truly seasonal breeders. Under natural lighting conditions, egg laying starts in early spring in temperate regions and in late autumn in subtropical areas when, depending on latitude, daylengths are between 10 and 12 h. Peak rate of lay is no more than 50% and production terminates after about 4 months of production; that is before the longest day (100, 101). However, egg production is markedly improved if daylength does not exceed 10 or 11 h, allowing the season to be extended and egg numbers to be significantly increased (Figure 3.35, 101, 102; Table 3.9, 101). This is because 8-h photoperiods are neutral, while those ≤11 h are only mildly stimulatory and the onset of adult photorefractoriness is delayed if birds are not exposed to a more stimulatory (≥ 12 h) daylength during the laying period. Annual egg production can also be increased by reducing the daylength to 7 h for about 8 weeks at midsummer, to dissipate photorefractoriness after the geese have completed their first egg laying cycle under natural lighting, then giving six 30-min daily increments to 10 h and maintaining this daylength for 20 to 22 weeks. This schedule can then be given repeatedly to create six laying cycles in 4 years (104).

Where lightproof housing is available, it is recommended that the daylength is reduced to 7 h for an 8-week period starting 12 weeks before the desired start of egg production, to dissipate juvenile photorefractoriness. Daily increments of 30 min to 10 or 11 h can then be given to initiate sexual development (104).

Sexual maturity in geese that had been transferred from 12-h to 14 or 18-h photoperiods at 19 weeks of age was significantly delayed (by 115 and 163 d, respectively),

Table 3.9 *Time to reach 40% lay, total egg production, mean egg weight and hatchability in geese maintained on 8, 10 or 12-h photoperiods from the shortest day (103).*

Trait	Photoperiod		
	8 h	10 h	12 h
40% lay (weeks)	9	5	4
Eggs/bird	60.0	62.4	55.3
Egg wt. (g)	179	178	174
Fertility (%)	90.5	92.4	91.6

Table 3.10 *Age at first egg (AFE±SEM), egg production, egg weight and fertility in geese transferred from 12-h photoperiods to natural (NAT) lighting (24°N) or to a constant artificial 14 or 18-h photoperiod at 19 weeks in late October (100).*

Trait	Photoperiod		
	NAT	14 h	18 h
AFE (d)	251±4	366±24	414±23
Eggs/bird	46.6	42.9	49.1
Egg wt. (g)	143.6	159.2	153.8
Fertility (%)	50.2	63.8	68.7

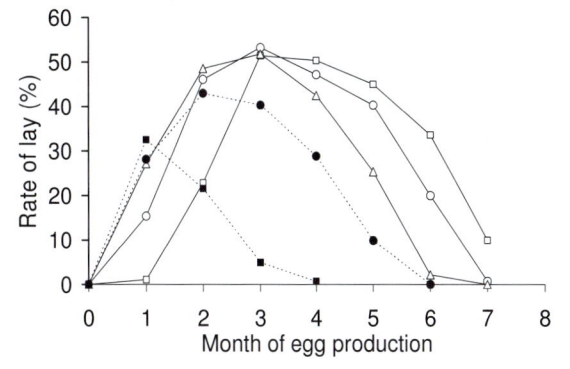

Figure 3.35 *Rate of lay in geese held on 11 (●) or 14-h (■), or 8 (□), 10 (○) or 12-h (△) photoperiods (102, 103).*

compared with naturally illuminated birds exposed to daylengths falling to 11.5 h at midwinter and then rising to 14.5 (Table 3.10). These data demonstrate that the rate at which juvenile photorefractoriness is dissipated in seasonal breeding species depends on daylength and, as in broiler breeders, birds reared on constant long-days show a wider range of individual ages at first egg (larger standard errors). Body weights were not reported, but it is not surprising that, as occurs in egg-type hybrids and broiler breeders, egg weight increased when maturity was delayed even though egg numbers were not significantly affected. Delaying maturity also enhanced fertility, though hatch of fertile eggs was unaffected (101). Once mature, egg production will be terminated and its resumption prevented by exposing geese to a 14.5-h daylength after the longest day.

Sexual maturation was accelerated when geese that naturally reproduce between December and May (northern hemisphere) were transferred in late November from natural lighting alone (11-h daylengths) to a 20-h photoperiod of natural and artificial light (105). Peak rate of lay was reached in the second month of lay, rather than in the fourth for naturally lit controls, but the birds spontaneously moulted after only 3 months egg production (Figure 3.36). Following a transfer back to natural lighting, the geese started a second laying cycle that lasted from May to July. Presumably the reduction from 20 h to natural daylengths (about 12 h in March) was sufficient to dissipate photorefractoriness in the experimental birds and make them responsive to the naturally increasing daylengths in April and May, even though the same daylengths were simultaneously inducing photorefractoriness in the controls. Although the total number of production days, annual egg production and mean fertility were not significantly different from the naturally illuminated controls, the markedly higher value of out-of-season hatching eggs and goslings in some parts of the world make these findings both economically important and biologically interesting.

Egg-laying and Meat-type Ducks

Ducks

There is no evidence that photoperiodic responses are different in egg-laying and meat strains of ducks, in contrast to egg-laying and meat-producing strains of chickens. However, modern strains of Pekin ducks, like broiler chickens, have been bred to produce offspring with much greater growth rates and potential mature body sizes than their 19th century ancestors and, as a consequence, reproductive performance is poor unless feed intake is strictly controlled during rearing. This means that, as with broiler breeders, onset of egg laying in modern Pekin breeding stock is primarily controlled by nutrition, and not by manipulation of photoperiod.

Wild mallard (*Anas platyrynchos*) are seasonal breeders and domesticated ducks exposed to natural changes in daylength also show clear evidence of seasonal effects, but the species does not exhibit juvenile photorefractoriness. Wild mallard maintained on long days (> 16 h) from hatching until after maturity will come into lay at about 22 weeks of age (106), whereas they do not begin to lay until 35 weeks if hatched in June at 52°N and raised in natural daylight (107). Domesticated Pekin ducks also mature normally if reared on constant long days, but they will lay sooner if given increasing daylengths during rearing.

Some of the potential effects of artificially manipulated photoperiods are illustrated in Figure 3.37. The major delay in sexual maturity for treatments 3 and 4, contrasted with 1 and 2, is the compounded result of holding the ducks on short days up to 24 weeks and simultaneously controlling growth by severe feed restriction; but the Figure also shows the retarding effect of placing ducklings on long days before returning them to short days at 8 weeks. Note that the later maturing groups had a higher peak rate of lay and they later

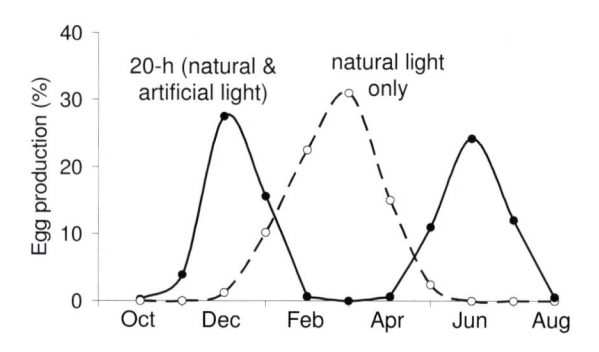

Figure 3.36 *Mean egg production for geese transferred from natural 11-h daylengths to 20 h of natural and artificial light at 20, 120 or 220 lux at the end of November and back to natural lighting after a moult in February (●), and for naturally illuminated controls (○) (105).*

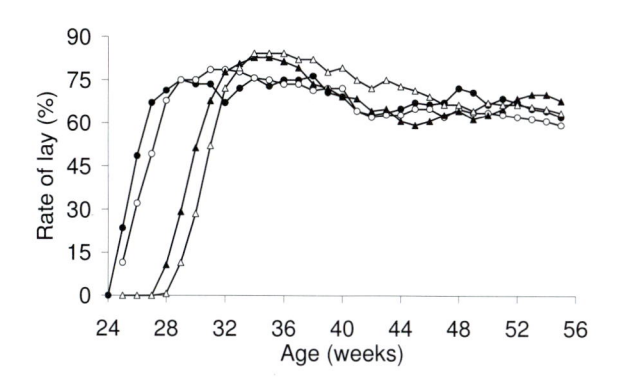

Figure 3.37 *Egg production for four groups of Pekin ducks exposed to different artificial light and feeding patterns during rearing. Groups 1 (●) and 3 (▲) were given 8 h photoperiods from hatching; groups 2 (○) and 4 (∆) had 23-h days from hatching until 8 weeks, when they were transferred to 8 h. All the ducks were held on 8-h days from 8 to 14 weeks. Groups 1 and 2 were then given Step Up lighting from 14 weeks to reach 17 h at 22 weeks and they were control-fed to reach the breeder's recommended target weight at 14 weeks; groups 3 and 4 had the same Step Up lighting, but not starting until 24 weeks, and their feed was more severely restricted so that they did not reach target body weight until 24 weeks (106).*

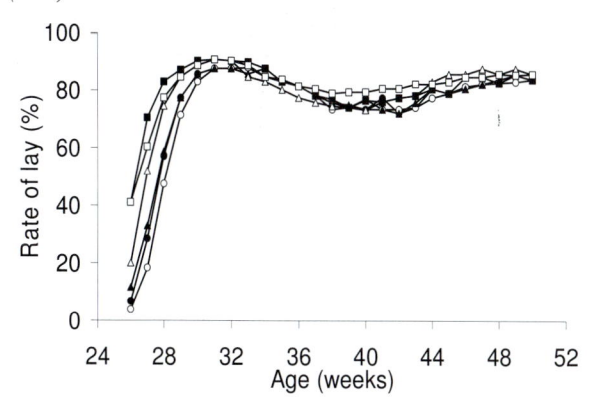

Figure 3.38 *Egg production for successive flocks of Pekin ducks hatched between June and December (●, ○, ▲, ∆, ■ and □) and exposed to natural daylight (52°N) from 5-18 weeks of age, then given Step-up lighting to reach 17 h at 26 weeks (106).*

tended to compensate for their 27 days' delay in maturity, but were still 17 eggs per bird behind the early maturing groups at 55 weeks of age.

Some of the effects of natural daylength changes on sexual maturity and subsequent laying performance are illustrated in Figure 3.38. The early maturity of the November (■) and

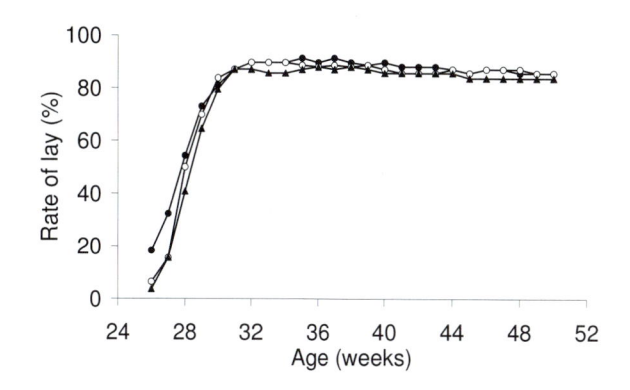

Figure 3.39 *Egg production for flocks of Pekin ducks hatched in March (●), April (○) and May (▲) and exposed to natural daylight (52°N) from 5-18 weeks, then given Step-up lighting to reach 17 h at 26 weeks (106).*

December (□) hatches, which had increasing daylengths throughout their development, contrasts with the later onset of lay in the June-August hatches that were exposed to shortening days until artificial lighting was applied at 18 weeks. Note that there is a post-peak depression in egg production in all these flocks and that it occurs at about the same stage of lay and therefore at differing dates, so that it cannot be correlated with prevailing environmental effects such as light, temperature, nutrition or disease. In contrast, flocks of the same genotype reared at the same location but hatched in March, April and May showed no post-peak depression in egg production (Figure 3.39). These latter groups experienced only small changes in natural daylength between 5 and 18 weeks and received correspondingly small increments in artificial daylength between 18 and 26 weeks. It has been proposed that the best way to treat breeding ducks, whether reared in lightproof houses or with exposure to variable natural days, is to give them a constant 17 h photoperiod from day-old to the end of their first laying season (106). Figure 3.40 shows that this is a highly successful treatment, at least for flocks that are subjected to feed restriction during rearing. Whether it would be equally successful with *ad libitum* fed ducks of egg-laying strains is not known.

Initial egg size is strongly correlated with age at first egg. In the trial illustrated in Figure 3.37, groups 1 & 2 had an average egg weight of 70 g

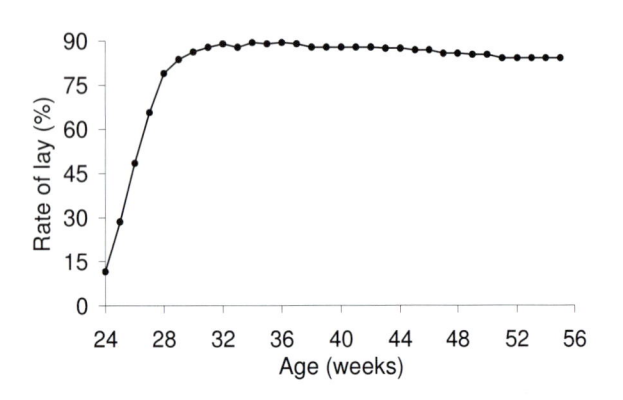

Figure 3.40 *Egg production for a flock of Pekin ducks given a constant 17-h day from hatching to 54 weeks of age (106).*

at 27-28 weeks of age, compared with 87 g for groups 3 & 4 at the same stage of lay (31-32 weeks). However, by 31-32 weeks, mean egg weight for the early maturing treatments had reached 86 g, showing that egg size is principally a function of age although, if feed intake is controlled during lay, this can also have an important influence on egg size. Given that eggs are not usually saved for incubation during the first few weeks of lay, there is no practically significant effect of lighting treatment upon hatching egg size.

Oviposition time

Time of lay in ducks, as in other bird species, is determined by the timing of ovulation, with an interval of 24 hours, or slightly more, between these two events. Ovulation is induced by a surge of LH secretion from the pituitary gland, which is in turn regulated by the time of sunset or lights off. In ducks, pre-ovulatory rises in plasma concentrations of LH and progesterone begin immediately after lights out (108), whereas in chickens these surges begin about 5 hours after lights out (109). As a result, ducks on 14 h or 16 h days lay their eggs just before or very soon after dawn, whereas chickens on the same photoperiods lay, on average, about 5 h later. The duck's habit of laying her eggs before going out to forage on the water presumably has survival value in the wild. Chickens, however, seem to have evolved to eat breakfast before returning to the nest to lay and some other species, such as the Japanese quail, lay late in the afternoon.

Drakes

Benoit demonstrated that immature drakes taken from natural winter daylengths will show dramatic testicular responses if exposed to increased photoperiods (see Figure 2.5, page 11). However, there is very little evidence about the effect of rearing photoperiod on the productive responses of modern meat-type drakes. The data in Table 3.11 show that satisfactory fertility can be obtained following a wide variety of rearing programmes applied to drakes. No doubt, rates of testicular development in these males would have been affected by the lighting treatments applied, but this did not influence their reproductive success at the time that hatching eggs were needed.

Table 3.11 *Fertility in ducks (all reared on a constant 17 h day) mated to drakes that had received a constant, a step-down or a step-up lighting regime during rearing (106).*

Age (weeks)	Constant 17 h from 0 to 55 weeks	Step down 23 h from 0 to 6 weeks, then reducing gradually to 17 h at 17 weeks	Step up 23 h from 0 to 6 weeks, dropping to 8 h at 6 weeks then increasing gradually to 17 h at 17 weeks
	(%)	(%)	(%)
28	90.6	90.5	88.5
29	90.4	91.0	91.6
30	92.1	93.7	92.0
31	92.9	93.0	92.5

Chapter 3 Photoperiod: Conventional programmes

• Dawn and dusk •

Although it is common for natural daylengths to be regarded as being from sunrise to sunset, activity observations in naturally illuminated laying hens suggest that their day begins 30 min before the start of morning civil twilight and ends 30 min before the end of evening civil twilight (110). The hen's day therefore has the same duration as the period between the beginning of morning and the end of evening twilight. The birds in this study got down from their perches when the illuminance was less than 0.03 lux, suggesting that they had anticipated daylight, but started perching when it was still 15 lux, presumably anticipating darkness, and whilst it was still adequately light for humans to perform outdoor tasks. The duration of civil twilight is dependent upon the distance from the equator, and so the bird's interpretation of day and night is doubly affected by latitude; by differences in sunrise and sunset, and the duration of the morning and evening twilight periods (Table 3.12, 111).

It is usual when poultry are artificially illuminated for the lights to be turned on abruptly at the beginning of the photoperiod and off at the end. However, there is some evidence that welfare and production benefits might accrue from the provision of a simulated dawn and dusk, especially the latter, and there is an EU statutory recommendation that laying hens are given a period of twilight before lights-out to facilitate roosting, especially when they are kept in alternative systems or enriched cages (112). It has even been suggested that, because dawn and dusk are such important periods for an animal, the failure to provide them might be stressful for the animal, particularly in the long-term (113).

Laying hens that were given a 5-min twilight period at the beginning and end of a 14-h photoperiod (24 lux) were observed to settle down for the night in a more orderly fashion than when the photoperiod started and ended abruptly (114). At the beginning of the dark period, virtually none had settled in the abrupt group whereas almost half were already sitting in the group given the twilight period. In contrast, hens in both groups had anticipated the photoperiod, and were standing ready to eat before the lights were turned on. In broiler chickens, however, there appears to be no change in perching behaviour in response to a 10-min simulated dusk (115).

The provision of 2-h morning and evening twilight periods within a 12-h day tends to result in higher feed intakes and faster growth in

Table 3.12 *Sunrise to sunset, morning and evening civil twilight, and poultry-perceived daylengths for the winter and summer solstice, according to latitude (111).*

	Winter solstice			Summer solstice		
Latitude (°N/°S)	Sunrise to sunset (h:min)	Morning and evening twilight (h:min)	Hen-perceived daylength (h:min)	Sunrise to sunset (h:min)	Morning and evening twilight (h:min)	Hen-perceived daylength (h:min)
0	12:08	0:44	12:52	12:07	0:45	12:52
5	11:50	0:45	12:35	12:25	0:45	13:10
10	11:32	0:46	12:18	12:42	0:47	13:29
15	11:14	0:47	12:01	13:01	0:47	13:59
20	10:56	0:48	11:44	13:20	0:50	14:21
25	10:35	0:50	11:25	13:42	0:50	14:32
30	10:13	0:53	11:06	14:05	0:55	15:00
35	9:48	0:56	10:44	14:31	0:59	15:30
40	9:20	1:01	10:21	15:01	1:06	16:07
45	8:46	1:08	9:54	15:37	1:15	16:52
50	8:04	1:18	9:22	16:22	1:30	17:52
55	7:10	1:31	8:41	17:22	1:57	19:19

broiler chicken, compared with a conventional 12-h photoperiod or continuous illumination (72). In particular, the dusk-stimulated feeding activity in the last two hours of the photoperiod, presumably in anticipation of the dark period, is valuable in ensuring that birds begin the night with full crops. The ability to anticipate night-time is thought to vary among individuals, and it is suggested that the provision of dusk enables all birds to perform crepuscular feeding, so preventing starvation during the night (116).

Most body-checked eggs, which are caused by a sudden muscular contraction in the shell gland when the shell is at a fragile stage of development, are laid early in the morning (28). The incidence is higher in hens given longer photoperiods, and this suggests that the stress-related adrenalin surge that triggers such a muscular contraction might be associated with the period when the birds are settling down for the night; and so the more orderly settling-down that occurs when hens are given an evening twilight might well ameliorate this problem.

• References •

1. Morris, T.R. (1967) Light requirements of the fowl, in: *Environmental Control in Poultry Production.* (Ed. T.C. Carter), pp. 15-39. Edinburgh, Oliver & Boyd.
2. Lewis, P.D., Morris, T.R. & Perry, G.C. (1998) A model for the effect of constant photoperiods on the rate of sexual maturation in pullets. *British Poultry Science* 39, 147-151.
3. Lewis, P.D. (1996) The domestic hen's response to photoperiodic influences, in: *Proceedings XX World's Poultry Congress, New Delhi, India.* Vol II, pp. 737-745.
4. King, D.F. (1961) Effects of increasing, decreasing and constant lighting treatments on growing pullets. *Poultry Science* 40, 479-484.
5. Lee, P.J.W., Gulliver, A.L. & Morris, T.R. (1971) A quantitative analysis of the literature concerning the restricted feeding of growing pullets. *British Poultry Science* 12, 413-437.
6. Gous, R.M. & Morris, T.R. (2001) The influence of pelleted feed on the response of growing pullets to photoperiods of less than ten hours. *British Poultry Science* 42, 203-206.
7. Lewis, P.D., Perry, G.C. & Morris, T.R. (1996) Effect of constant and of changing photoperiods on age at first egg and related traits in pullets. *British Poultry Science* 37, 885-894.
8. Sharp, P.J. (1993) Photoprerioidic control of reproduction in the domestic fowl. *Poultry Science* 72, 897-905.
9. Lewis, P.D. & Morris, T.R. (2005) Change in the effect of constant photoperiods on the rate of sexual maturation in modern genotypes of domestic pullet. *British Poultry Science*.46, 584-586.
10. Havenstein, G.B., Ferket, P.R., Scheideler, S.E. & Larson, B.T. (1994) Growth, liveability, and feed conversion of 1957 vs 1991 broilers fed 'typical" 1957 and 1991 diets. *Poultry Science* 73, 1785-1794.
11. Morris, T.R. (1980) Recent developments in lighting patterns for poultry in light-proof housing. *Proceedings of South Pacific Poultry Science Convention*, Auckland, New Zealand, pp. 116-124.
12. Leeson, S., Caston, L. & Lewis, P.D. (2005) Rearing and laying performance following various step-down lighting regimens in the rearing period. *Poultry Science* 84, 626-632.
13. Lewis, P.D., Morris, T.R. & Perry, G.C. (2002) A model for predicting the age at sexual maturity for growing pullets of layer strains given a single change in photoperiod. *Journal of Agricultural Science, Cambridge* 138, 441-458.
14. Dunn, I.C. & Sharp, P.J. (1990) Photoperiodic requirements for LH release in juvenile broiler and egg-laying strains of domestic chickens fed ad libitum or restricted diets. *Journal of Reproduction and Fertility* 90, 329-335.

15. Lewis, P.D., Perry, G.C., Morris, T.R., Douthwaite, J.A. & Bentley, G.E. (1998) Effect of constant and of changing photoperiod on plasma LH and FSH concentrations and age at first egg in layer strains of domestic pullets. *British Poultry Science* 39, 662-670.
16. Dunn, I.C., Lewis, P.D., Wilson, P.W. & Sharp, P.J. (2003) Oestrogen accelerates the maturation of FSH and LH responses to photostimulation in prepubertal domestic hens. *Reproduction* 126, 217-225.
17. Lewis, P.D., Dunn, I.C., Perry, G.C., Morris, T.R. & Sharp, P.J. (2001) Effect of exogenous oestradiol and age at photostimulation on age at first egg in domestic pullets. *British Poultry Science* 42, 530-535.
18. Lewis, P.D., Morris, T.R. & Perry, G.C. (2003) Effect of two opposing changes in photoperiod upon age at first egg in layer strains of pullets. *Journal of Agricultural Science, Cambridge* 140, 373-379.
19. Lewis, P.D. & Gous, R.M. (2004) Effect of one or two pre-pubertal long-days on age at first egg in domestic pullets. *British Poultry Science,* 45, 28-30.
20. Lewis, P.D., Perry, G.C. & Morris, T.R. (1996) Effects of 5 h increases in photoperiod and in feeding opportunity on age at first egg. *British Poultry Science* 37, 15-19.
21. Goodale, H.D. (1918) Internal factors influencing egg production in the Rhode Island Red breed of domestic fowl. *The American Naturalist* LII, part I 65-94, part II 209-232, part III 301-321.
22. Morris, T.R. & Fox, S. (1958) Light and sexual maturity in the domestic fowl. *Nature* 181, 1453-1454.
23. Lewis, P.D., Perry, G.C. & Morris, T.R. (1994) Lighting and egg shell quality. *World's Poultry Science Journal* 50, 288-291.
24. Klandorf, H., Boyce, C.S., Killefer, J., McGowan, J.A., Petersen, R.A., Valent, M. & Deaver, D.R. (1997) The effect of photoperiod and food intake on daily changes in plasma calcitonin in broiler breeder hens. *General and Comparative Endocrinology* 107, 327-340.
25. Ieda, T., Takahashi, T., Saito, N., Yasuoka, T., Kawashima, M. & Shimada, K. (2000) Changes in parathyroid hormone-related peptide receptor binding in the shell gland of laying hens (Gallus domesticus) during the oviposition cycle. *General and Comparative Endocrinology* 117, 182-188.
26. Ostrowska, Z., Kos-Kudla, B., Marek, B. & Kajdaniuk, D. (2003) Influence of lighting conditions on daily rhythm of bone metabolism in rats and possible involvement of melatonin and other hormones in this process. *Endocrine Regulations* 37, 163-174.
27. Cardinali, D.P., Ladizesky, M.G., Boggio, V., Cutera, R.A. & Mautalen, C. (2003) Melatonin effects on bone: experimental facts and clinical perspectives. *Journal of Pineal Research* 34, 81-87.
28. Roland, D.A. & Moore, C.H. (1980) Effect of photoperiod on the incidence of body-checked and misshapen eggs. *Poultry Science* 59, 2703-2707.
29. DEFRA (2002) Environment, in: *Code of recommendations for the welfare of livestock – laying hens.* p. 17. London, DEFRA Publications.
30. Lewis, P.D., MacLeod, M.G. & Perry, G.C. (1994) Effects of lighting regime and grower diet energy concentration on energy expenditure, fat deposition, and body weight gain of laying hens. *British Poultry Science* 35, 407-415.
31. Berman, A. & Meltzer, A. (1978) Metabolic rate: its circadian rhythmicity in the female domestic fowl. *Journal of Physiology, London* 282, 419-427.
32. Perry, G.C. & Lewis, P.D. (1992) Feeding activity of laying hens under 8, 11 and 14 hour photoperiods. *Proc. XIX World's Poultry Congress, Amsterdam, The Netherlands.* Vol 3, 619-622.
33. Mongin, P. & Sauveur, B. (1974) Voluntary food and calcium intake by the laying hen. *British Poultry Science* 15, 349-359.
34. Savory, C.J. (1976) Effects of different lighting regimes on diurnal feeding patterns of the domestic fowl. *British Poultry Science* 17, 341-350.
35. Lewis, P.D., Morris, T.R. & Perry, G.C. (1996) Lighting and mortality rates in domestic fowl. *British Poultry Science* 37, 295-300.
36. Wilson, S.C. & Cunningham, F.J. (1984) Endocrine control of the ovulation cycle, in: *Reproductive biology of poultry* (Eds. F.J. Cunningham, P.E. Lake & D. Hewitt). pp. 29-49, Harlow, Longman.

37. Bhatti, B.M. & Morris, T.R. (1988) Model for the prediction of mean time of oviposition for hens kept in different light and dark cycles. *British Poultry Science* 29, 205-213.
38. Lewis, P.D., Perry, G.C. & Morris, T.R. (1995) Effect of photoperiod on the mean oviposition time of two breeds of laying hen. *British Poultry Science* 36, 33-37.
39. Nøddegaard, F. (1996) Role of melatonin in chicken egg laying cycles. *PhD thesis*, Royal Veterinary and Agricultural University, Denmark.
40. Gow, C.B., Sharp, P.J., Carter, N.B., Scaramuzzi, R.J., Sheldon, B.L., Yoo, B.H. & Talbot, R.T. (1985) Effects of selection for reduced oviposition interval on plasma concentrations of luteinising hormone during the ovulatory cycle in hens on a 24 h lighting cycle. *British Poultry Science* 26, 441-451.
41. Lewis, P.D., Backhouse, D. & Gous, R.M. (2004) Photoperiod and oviposition time in broiler breeders. *British Poultry Science* 45, 561-564.
42. King, D.F. (1959) Artificial light for growing and laying birds. *Progress Report 72*, Agricultural Experiment Station, Alabama Polytechnic Institute, Auburn, USA.
43. Sharp, P.J. (1993) Photoperiodic control of reproduction in the domestic hen. *Poultry Science* 72, 897-905.
44. Lewis, P.D., Perry, G.C. & Morris, T.R. (1996) Effects of changes in photoperiod and feeding opportunity on the performance of two breeds of laying hen. *British Poultry Science* 37, 279-293.
45. Morris, T.R., Fox, S. & Jennings, R.C. (1964) The response of laying pullets to abrupt changes in daylength. *British Poultry Science* 5, 133-147.
46. Lewis, P.D. & Perry, G.C. (1990) Responses of the laying hen to noise supplemented short daylengths. *Proceedings VIII European Poultry Conference*, Barcelona, Spain, pp. 642-645.
47. Sharp, P.J., Dunn, I.C. & Cerolini, S. (1992) Neuroendocrine control of reduced persistence of egg-laying in domestic hens: evidence for the development of photorefractoriness. *Journal of Reproduction & Fertility* 94, 221-235.
48. Morris, T.R., Sharp, P.J. & Butler, E.A. (1995) A test for photorefractoriness in high-producing stocks of laying pullets. *British Poultry Science* 36, 763-769.
49. Lewis, P.D., Backhouse, D. & Gous, R.M. (2004) Constant photoperiods and sexual maturity in broiler breeder pullets. *British Poultry Science* 45, 557-560.
50. Farner, D.S. & Follett, B.K. (1966) Light and other environmental factors affecting avian reproduction. *Journal of Animal Science* 25 (Suppl), 90-115.
51. Renden, J.A., Oates, S.S. & West, M.S. (1991) Performance of two male broiler breeder strains raised and maintained on various constant photoschedules. *Poultry Science* 70, 1602-1609.
52. Lewis, P.D., Ciacciariello, M. & Gous, R.M. (2003) Photorefractoriness in broiler breeders: sexual maturity and egg production evidence. *British Poultry Science* 44, 634-642.
53. Lewis, P.D., Backhouse, D. & Gous, R.M. (2005) Effects of photoperiod and body weight on laying performance in broiler breeder domestic fowl. *Journal of Agricultural Science, Cambridge* 143, 97-108.
54. MacLeod, M.G., Jewitt, T.R. & Anderson, J.E.M. (1988) Energy expenditure and physical activity in domestic fowl kept on standard and interrupted lighting patterns. *British Poultry Science* 29, 231-244.
55. Lewis, P.D., Backhouse, D. & Gous, R.M. (2004) Photoperiod and oviposition time in broiler breeders. *British Poultry Science* 45, 561-564.
56. Lewis, P.D. (1987) Responses of laying hens to interrupted lighting regimes. *PhD thesis*, University of Bristol, UK.
57. Etches, R.J. (1996) The ovary, in: *Reproduction in poultry*, pp. 125-166. Wallingford, CAB International.
58. Backhouse, D., Lewis, P.D. & Gous, R.M. (2005) Constant photoperiods and eggshell quality in broiler breeders. *British Poultry Science* 46, 211-213.
59. Roque, L. & Soares, M.C. (1994) Effects of eggshell quality and broiler breeder age on hatchability. *Poultry Science* 73, 1838-1845.

60. Dunn, I.C. & Sharp, P.J. (1992) The effect of photoperiodic history on egg laying in dwarf broiler hens. *Poultry Science* 71, 2090-2098.
61. Gous, R.M. & Cherry, P. (2004) Effects of body weight at, and lighting regimen and growth curve to, 20 weeks on laying performance in broiler breeders. *British Poultry Science* 45, 445-452.
62. Lewis, P.D. & Gous, R.M. (2006) Various photoperiods and *Biomittent*™ lighting during rearing for broiler breeders subsequently transferred to open-sided housing at 20 weeks. *British Poultry Science* in press.
63. Lewis, P.D. & Gous, R.M. (2006) Effect of final photoperiod and 20-week body weight on sexual maturity and egg production in broiler breeders. *Poultry Science* in press.
64. Lien, R.J. & Yuan, T. (1994) Effect of delayed light stimulation on egg production by broiler breeder pullets of low body weight. *Applied Poultry Science* 3, 40-48.
65. Joseph, N. S., F. E. Robinson, R. A. Renema, and M. J. Zuidhof. (2002) Responses of two strains of female broiler breeders to a midcycle increase in photoperiod. *Poultry Science* 81:745-754.
66. Vandenberghe, N., Ceular, A.L. & Moreno, A. (1999) Use of hatch date for broiler breeder production planning. *Poultry Science* 78, 501-504.
67. Lopez, G. & Leeson, S. (1992) Manage broiler breeder pullets to reduce seasonal variations. *Poultry Digest* September, 31-35.
68. Cherry, P. & Barwick, M.W. (1962) The effect of light on broiler growth. II. Light patterns. *British Poultry Science* 3, 41-50.
69. Renden, J.A., Bilgili, S.F. & Kincaid, S.A. (1993) Comparison of restricted and increasing light programs for male broiler performance and carcass yield. *Poultry Science* 72, 378-382.
70. Lewis, P.D. (2001) Lighting regimes for broiler and egg production, in: *Proceedings of XVII Latin American Poultry Congress*, pp. 326-335.
71. Savory, C.J. (1976) Broiler growth and feeding behaviour in three different lighting regimes. *British Poultry Science* 17, 557-560.
72. Classen, H.L. & Riddell, C. (1989) Photoperiodic effects on performance and leg abnormalities in broiler chickens. *Poultry Science* 68, 873-879.
73. Classen, H.L., Riddell, C. & Robinson, F.E. (1991) Effects of increasing photoperiod length on performance and health of broiler chickens. *British Poultry Science* 32, 21-29.
74. Wise, D.R. (1970) Comparisons of the skeletal systems of growing broiler and layer strain chickens. *British Poultry Science* 11, 333-339.
75. Thorp, B.H. & Duff, S.R. (1988) Effect of exercise on the vascular pattern in the bone extremities of broiler fowl. *Research in Veterinary Science* 45, 72-77.
76. Stanley, V.G., Gutierrez, J., Parks, A.L., Rhoden, S.A., Chukwu, H., Gray, C. & Krueger, W.F. (1997) Relationship between age of commercial broiler chickens and response to photostimulation. *Poultry Science* 76, 306-310.
77. Lewis, P.D., Perry, G.C. & Sherwin, C.M. (1998) Effect of photoperiod and light intensity on the performance of intact male turkeys. *Animal Science* 66, 759-767.
78. Siopes, T.D., Baughman, G.R., Parkhurst, C.R. & Timmons, M.B. (1989) Relationship between duration and intensity of environmental light on the growth performance of male turkeys. *Poultry Science* 68, 1428-1435.
79. Newberry, R.C. (1992) Influence of increasing photoperiod and toe clipping on breast buttons of turkeys. *Poultry Science* 71, 1471-1479.
80. Siopes, T.D., Baughman, G.R. & Parkhurst, C.R. (1993) Photoperiod and seasonal influences on the growth of turkey hens. *British Poultry Science* 34, 43-51.
81. Hester, P.Y., Elkin, R.G. & Klingensmith, P.M. (1983) Effects of high intensity step-up and low intensity step-down lighting programs on the incidence of leg abnormalities in turkeys. *Poultry Science* 62, 887-896.
82. Hester, P.Y., Sutton, A.L.., Elkin, R.G. & Klingensmith, P.M. (1985) The effect of lighting, dietary amino acids, and litter on the incidence of leg abnormalities and performance of turkey toms. *Poultry Science* 64, 2062-2075.

83. Lilburn, M.S., Renner, P.A. & Anthony, N.B. (1992) Interaction between step-up versus step-down lighting from four to sixteen weeks on growth and development in turkey hens from two commercial strains. *Poultry Science* 71, 419-426.
84. Wilson, W.O., Ogasawara, F.X. & Asmundson, V.S. (1962) Artificial control of egg production in turkeys by photoperiods. *Poultry Science* 41, 1168-1175.
85. Siopes, T.D. (1994) Critical day lengths for egg production and photorefractoriness in the domestic turkey. *Poultry Science* 73, 1906-1913.
86. Lien, R.J. & Siopes, T.D. (1993) The relationship of plasma thyroid hormone and prolactin concentrations to egg laying, incubation behaviour, and molting by female turkeys exposed to a one-year natural daylength cycle. *General and Comparative Endocrinology* 90, 205-213.
87. Siopes, T.D. (1998) Absence of a seasonal effect on the critical day length for photorefractoriness in turkey breeder hens. *Poultry Science* 77, 145-149.
88. McCartney, M.G., Sanger, V.L., Brown, K.I. & Chamberlin, V.D. (1961) Photoperiodism as a factor in the reproduction of the turkey. *Poultry Science* 40, 368-376.
89. Lewis, P.D. & Morris, T.R. (1998) A comparison of the effects of age at photostimulation on sexual maturity and egg production in domestic fowl, turkeys, partridges and quail. *World's Poultry Science Journal* 54, 119-128.
90. Woodard, A.E. & Snyder, R.L. (1976) Sexual maturity and persistency of lay in the Chukar Partridge given stimulatory light at different ages. *Poultry Science* 55, 2461-2463.
91. Stein, G.S. & Bacon, W.L. (1976) Effect of photoperiod upon age and maintenance of sexual development in female *Coturnix coturnix japonica*. *Poultry Science* 55, 1214-1218.
92. Lewis, P.D., Perry, G.C., & Morris, T.R. (1997) Effect of size and timing of photoperiod increase on age at first egg and subsequent performance of two breeds of laying hen. *British Poultry Science* 38, 142-150.
93. Gueme, D. (1990) Effect of changing lighting programs at various ages on laying performance in turkey hens (*Meleagris gallopavo*) housed in cages or in floor pen, in: *Control of fertility in domestic birds*. pp. 233-241. Paris, INRA.
94. Jacquet, J.M. & Sauveur, B. (1995) Photoperiodic control of sexual maturation in muscovy drakes. *Domestic Animal Endocrinology* 12, 189-195.
95. Sauveur, B. & de Carville, H. (1990) Contrôle de la précocité sexuelle de la cane de Barbarie par la photopériode, in: *Control of fertility in domestic birds* (Ed. J. Brillard). pp. 197-203. Paris, INRA.
96. Jacquet, J.M. (1997) Photorefractory period of the muscovy duck (*Cairina moschata*): endocrine and neuroendocrine responses to day length after a full reproductive cycle. *British Poultry Science* 38, 209-216.
97. Siopes, T.D. (2001) Temporal characteristics and incidence of photo-refractoriness in turkey hens. *Poultry Science* 80, 95-100.
98. Jallageas, M., Tamisier, A. & Assenmacher, I. (1978) A comparative study of the annual cycles in sexual and thyroid function in male Pekin ducks (*Anas platyrhyncos*) and teal (*Anas crecca*). *General and Comparative Endocrinology* 36, 201-210.
99. Hasse, E. & Paulke, E. (1980) Plasma concentrations of tri-iodothyronine and testosterone during the annual cycle of wild mallard drakes. *Zoologischer Anzeiger (Jena)* 204S, 102-110.
100. Gillette, D.D. (1976) Laying patterns of geese in the Mid West. *Poultry Science* 55, 1143-1146.
101. Wang, S.D., Jan, D.F., Yeh, L.T., Wu, G.C. & Chen, L.R. (2002) Effect of exposure to long photoperiod during the rearing period on the age at first egg and the subsequent reproductive performance in geese. *Animal Reproduction Science* 73, 227-234.
102. Rousselot-Pailley, D.T. & Sellier, N. (1990) Influence de quelque facteurs zootechniques sur la fertilitie des oies, in: *Control of fertility in domestic birds* (Ed. J. Brillard). pp. 145-154, Paris, INRA.
103. Elminowska-Wenda, G. & Rosiński, A. (1993) Wpyw długości dnia świetlnego na wyniki rozrodu gęsi białych włoskich. *Roczniki Naukowe Zootechniki* 20, 339-351.
104. FAO. (2002) Breeder flock management, in: *Goose production: FAO Animal production and health paper 154* (Ed. R. Buckland). pp. 23-35. Rome, FAO.

105. Wang, C.M., Kao, J.Y., Lee, S.R. & Chen, L.R. (2005) Effects of artificial supplemental light on reproductive season of geese kept in open houses. *British Poultry Science* 46, 728-732.
106. Cherry, P. (1993) *Sexual Maturity in the Domestic Duck*. PhD thesis, University of Reading.
107. Hill, D.A. (1984) Laying date, clutch size and egg size of the Mallard, Anas platyrhyncos, and the Tufted duck, Aythya fuligula. *Ibis* 126, 484-495.
108. Tanabe, Y., Nakamura, T., Omiya, Y. & Yano, T. (1980) Changes in the plasma LH, progesterone and estradiol during the ovulatory cycle of the duck (*Anas platyrhynchos*) exposed to different photoperiods. *General and Comparative Endocrinology* 41, 378-383.
109. Williams, J.B. & Sharp, P.J. (1978) Control of preovulatory surge of luteinizing hormone in the hen (*Gallus domesticus*): the role of progesterone and androgens. *Journal of Endocrinology* 77, 57-65.
110. Yeates, N.T.M. (1963) The activity pattern in poultry in relation to photoperiod. *Animal Behaviour* 11, 287-289.
111. Astronomical Applications Department, US Naval Observatory, *via* Royal Observatory, Edinburgh website: www.roe.ac.uk/info/srss.html.
112. DEFRA (2002) Environment, in: *Code of recommendations for the welfare of livestock – laying hens*. pp. 17-18, London, DEFRA publications.
113. Bryant, S.L. (1987) A case for dawn and dusk for housed livestock. *Applied Animal Behavioural Science* 18, 379-382.
114. Tanaka, T. & Hurnik, J.F. (1991) Behavioral responses of hens to simulated dawn and dusk periods. *Poultry Science* 70, 483-488.
115. Martrenchar, A., Huonnic, D., Cotte, J.P., Boilletot, E. & Morisse, J.P. (2000) Influence of stocking density, artificial dusk and group size on the perching behaviour of broilers. *British Poultry Science* 41, 125-130.
116. Savory, C.J. (1980) Diurnal feeding patterns in domestic fowl: a review. *Applied Animal Ethology* 6, 71-82.

4 PHOTOPERIOD
Unconventional programmes

This chapter describes the bird's responses to unconventional light-dark cycles. These include intermittent programmes, which have more than one light and dark period in each 24-h cycle, ahemeral cycles, where there is one light and dark period per cycle but the cycle length is either less or more than 24 h, and the constant conditions of continuous light or darkness.

• Intermittent programmes •

Intermittent lighting programmes have variously been described as restricted, interrupted, short-cycle, skeletal, segmented, night interruption, flash or *Biomittent®* - a term registered by Ralston Purina Company, St. Louis, U.S.A.

Intermittent lighting was first used to study the avian photoperiodic response; ranging from the minimum amount of light required to sustain egg production (1, 2) to identification of the photoinducible phase (3). Subsequently, the emphasis changed from pure to applied research, with studies conducted to exploit the economic advantages of programmes that achieved equal or better performance when compared with conventional lighting.

Intermittent programmes can be classified into two main types according to their ability to synchronise circadian rhythms (Table 4.1) (4).

Asymmetrical regimens

These programmes generally have one scotoperiod within the 24-h cycle that is perceptibly longer than any of the others and which is regarded by the birds as night, whilst the remainder of the 24 h, whether the lights are on or off, is perceived as a subjective day, with all birds in a group treating the light-dark interface at the beginning of the night as dusk (5). For example, 8L:4D:2L:10D (L=light, D=darkness) and 13(0.5L:0.5D):0.5D:0.5L:10D

Table 4.1 *Classification and characteristics of intermittent lighting regimens.*

Regimen names	Examples (L=light : D=darkness, h)	Perceived regimen (L=light : D=darkness, h)	Synchronising effect
Asymmetrical			
Cornell	2L:4D:8L:10D	14L:10D	Synchronised
Biomittent®	15(0.25L:0.75D): (0.25L:0.5D:0.25L):8D	16L:8D	Synchronised
Flash or night interruption	15.5L:5D:0.5L:3D	19L:5D	Synchronised
Symmetrical			
Short-cycle	4(3L:3D) or 3(2.5L:5.5D)	Constant conditions	Free running
Long-cycle	2(2L:10D)	14L:10D	Synchronised
Reading	24(0.25L:0.75D)	Constant conditions	Free running

® registered by Ralston Purina Company, USA.

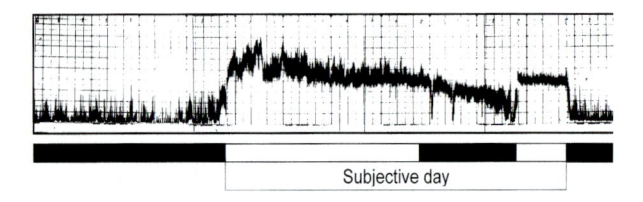

Figure 4.1 *Noise output for laying hens on 8L:4D:2L:10D asymmetrical lighting regimen (white bars=photoperiod, black bars= scotoperiod) (6).*

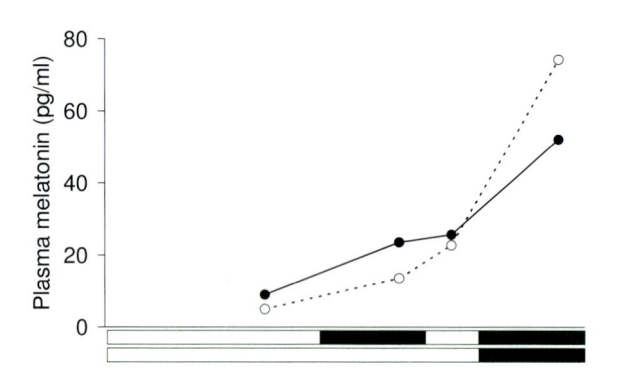

Figure 4.2 *Plasma melatonin concentration in laying hens on 8L:4D:2L:10D (○) or 14L:10D (●) lighting regimen. (white bars=photoperiod, black bars= scotoperiod) (7).*

are both interpreted as a 14-h day and a 10-h night. Noise output from laying hens on an 8L:4D:2L:10D regimen clearly shows that the 4-h scotoperiod is perceived as being part of the day and not the night (Figure 4.1, 6). However, there is a marked increase in noise at the beginning of the 2-h photoperiod in which concentrated crepuscular feeding results in a consistent noise output. Plasma melatonin concentrations also confirm that the shorter scotoperiod combines with the two photoperiods to form a 14-h subjective day (Figure 4.2, 7). Where the programme has equal-size scotoperiods, but one of the photoperiods is substantially longer than the other, for example, 14L:4D:2L:4D, the bird is still able to perceive a day and night with the day starting at the beginning of the long photoperiod (8).

Biomittent lighting is a particular type of asymmetrical lighting in which each hour of the subjective day (usually 16 h) is nominally segmented into 15 min light and 45 min darkness. It was developed for laying hens by the Ralston Purina Company in America to achieve greater savings in feed and electricity than accrue with the Cornell programme (9).

Circadian rhythms, such as the ovulatory cycle, feeding pattern, and activity are synchronised under asymmetrical lighting programmes because all birds are able to form a subjective day and night and have a common dusk for phase-setting of these rhythms. This results in complete entrainment of egg-laying and a mean oviposition time (MOT) that is similar to that for hens on the equivalent conventional photoperiod (10). For example, if hens on 14L:10D have an MOT of 10.30, then so will hens exposed to either 8L:4D:2L:10D or *Biomittent* lighting, with all groups using the light-dark interface at the beginning of the 10-h night as the zeitgeber for fixing the position of the open-period for pre-ovulatory luteinizing hormone release.

Symmetrical regimens

In symmetrical programmes, all the light and dark periods are the same size, though not necessarily equal to each other, for example 4(3L:3D) or 4(1.5L:4.5D). When all the scotoperiods are short and of the same length, the birds cannot discern a day and night and so they free-run as if in continuous illumination. Under these conditions, the birds function under the influence of an internal clock that oscillates with a frequency of about 25.3-h (11). This occurs even though the light-dark cycles fit into 24 h exactly. In symmetrical regimens with longer photoperiods, e.g. 2(2L:10D), the bird is able to form a subjective day by combining the two photoperiods and one of the scotoperiods and is thus synchronised to a 24-h cycle.

Night interruption

A special type of intermittent programme is night interruption or flash lighting. A light pulse of as little as 20 s but more usually about 1 or 2 h is given to birds that have previously been exposed to a conventional regimen. In the 1950s, flash lighting was used for commercial laying hens to stimulate egg production during short winter days. This involved the provision of three or four 20-s flashes from powerful 1500W lamps at hourly intervals in the middle of the night (12). A single light pulse, but at normal illuminance, is commonly used in photoperiodic research to identify the location of the photoinducible phase (see pages 13-14).

Which intermittent type to use?
The type of intermittent lighting programme used and the age at which it is introduced depends on what advantage a producer requires and, in some parts of the world, the need to comply with statutory welfare regulations.

Growing pullets
Intermittent lighting programmes are rarely used during the rearing period for growing pullets because (a), the pullets are usually on a short photoperiod, so there is little opportunity to save electricity and (b), lighting programmes for modern hybrids frequently need to maximise rather than minimise feed intake for the birds to reach recommended body weight targets.

When pullets were reared on a 4(1.5L:4.5D) regimen from hatch and progressively transferred to a 4(3L:3D) schedule from 18 weeks of age, sexual maturity, egg production and mean egg weight to 60 weeks were reported to be similar to that for birds reared on conventional 8-h photoperiods and transferred to 4(3L:3D) lighting at 20 weeks (13). However, the intermittently reared birds laid eggs with significantly heavier shells and greater breaking strength, but had higher feed intake and poorer feed conversion efficiency than the conventionally reared pullets.

Laying hens
Asymmetrical programmes
An important asymmetrical programme (2L:4D:8L:10D) was developed at Cornell University (3). Hens subjected to this type of lighting generally produce the same egg output, both in terms of numbers and weight, as birds on the equivalent solid photoperiod regime, eg., birds on 8L:4D:2L:10D perform similarly to those on 14L:10D (Table 4.2, 14). However, rate of lay is inferior, though egg weight unaffected, when the total amount of illumination is less than 4 h per day (15) or when the regimen is perceived as a very long day, e.g. 14L:4D:2L:4D which is treated as a 20-h subjective day (8). Savings in feed intake and improvements in feed conversion efficiency are regularly achieved but invariably fail to be statistically significant. However, the consistency of the reduction in feed intake (mean saving of 2.0% for 19 of the 25 comparisons in Table 4.2 and 0.8% increase for the remaining 6 trials) and commercial experience suggest that the savings are real. Part of the reduction in intake is likely to be a consequence of the reduced energy expenditure in an intermittent schedule; conventionally illuminated hens expend about 1% less energy per hour in darkness than in light (16). A study with laying hens on a 4(1L:3.5D):1L:5D schedule reported that mean daily oxygen usage was 15% lower than for normally illuminated 14-h controls (17). Additionally, laying hens given a 14-h 'day' interrupted by a 4-h dark period located 2, 6 or 8 h after dawn conducted 26% less daily feeding activity than 14L:10D controls (18).

Biomittent lighting
To acclimatise birds to *Biomittent* lighting, it is normal to introduce the programme gradually by segmenting the hours into 0.75L:0.25D, then to 0.5L:0.5D, and finally to 0.25L:0.75D. However, it is possible to introduce it abruptly provided the diet is adequately formulated for the reduced feed intake (18). The last hour of a *Biomittent* programme is commonly segmented as 15min light, 30 min dark, 15 min light to create a subjective day that has a whole number of hours, otherwise a 16(0.25L:0.75D):8D schedule would be perceived as 15.25 h rather than 16 h.

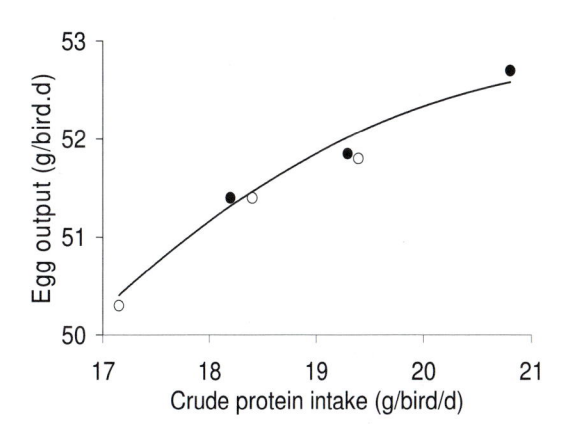

Figure 4.3 *Mean daily egg output between 38 and 72 weeks for laying hens on Biomittent lighting (○) or a conventional programme stepping up to 16 h at 41 weeks (●) and receiving three levels of dietary protein (21).*

Table 4.2 *Comparative responses of laying hens to asymmetrical, Biomittent, short-cycle, long-cycle and the Reading System of intermittent lighting regimens (conventionally lighted controls =100). Mean ±SEM (number of comparisons in parentheses) (14, 19, 21, 22, 23, 24, 25).*

Trait	Asymmetrical	*Biomittent*[1]	Short-cycles	Long-cycles	Reading System[2]
Egg numbers	100.3±0.4 (45)	98.9±0.8 (7)	95.6±0.7 (18)	95.3±1.6 (8)	96.7 (1)
Mean egg weight	100.4±0.2 (34)	99.9±0.1 (7)	102.2±0.4 (18)	101.5±0.3 (6)	102.0 (1)
Total egg mass	100.7±0.6 (34)	98.8±0.8 (7)	97.6±0.5 (18)	96.2±1.9 (6)	98.7 (1)
Feed intake	98.6±0.3 (26)	96.3±0.5 (7)	96.4±0.7 (18)	96.1±0.6 (8)	94.2 (1)
Feed conversion efficiency	101.4±0.5 (25)	102.6±0.9 (7)	101.2±0.9 (18)	100.2±2.1 (6)	104.7 (1)
Shell quality	102.4±1.0 (21)	102.0±0.5 (3)	104.9±1.1 (13)	105.3±0.9 (4)	103.0 (1)
End of lay body weight	98.8±1.4 (8)	97.9±2.4 (3)	103.2±1.5 (6)	102.3±2.3 (2)	-

[1] Introduced after peak rate of lay
[2] Continuously repeating 15 min light, 45 min dark cycle

Egg production reported for hens subjected to *Biomittent* lighting has been variable and depends, among other things, on the age at which the programme is introduced and the nutrient concentration of the diet. If pullets are transferred to *Biomittent* lighting before 24 weeks, may be peak rate of lay is adversely affected if feed savings are excessive or the birds are given an inadequately formulated diet (9, 20). When the programme is introduced after peak egg production or when the diet is satisfactory for the level of feed intake (19, 21), egg output is not significantly different from that of conventionally illuminated controls. *Biomittent* consistently yields bigger feed savings than the Cornell programme (Table 4.2), because the reduced amount of illumination leads to a greater reduction in feeding activity (18). Thus, when egg output (usually egg weight) is depressed it is most likely the result of a suboptimal nutrient intake; for a given intake of protein, egg output is the same for *Biomittent* and conventional controls (Figure 4.3, 21).

Hens given asymmetrical lighting have a 2% lower body weight at end of lay (Table 4.2).

Short-cycles

Short-cycle regimens were developed in France, and were found to increase egg weight and improve shell quality (13). Laying hens on short cycles such as 4(3L:3D) are unable to form a day and night and, as a consequence, are released from the confines of the 24-h solar day. This free-running results in an extension of the ovulatory cycle, the egg spending longer in the oviduct and, as a result, having a thicker shell. The longer interval between ovulations results in larger yolks, giving eggs that are 1-2 g heavier than conventionally lighted controls (Table 4.2). However, an adverse consequence of the longer ovulatory cycle is a reduction of about 4% in egg numbers, although this is smaller when hens are transferred to a 3L:3D regimen later in the laying-year, when the mean intra-sequence interval between ovipositions exceeds the 25.3-h internal circadian rhythmicity.

The inability to form a day and night and the lack of a common 'dusk' to synchronise the setting of the open-period for pre-ovulatory LH release result in eggs being laid randomly throughout the 24 h when a flock of hens is transferred to 3L:3D symmetrical lighting in lightproof accommodation before they have reached sexual maturity. However, egg-laying is more concentrated during the first half of the solar day when there is disparity between the light and dark periods, e.g. a 1.5L:4.5D schedule (13), and an almost complete synchrony of egg-laying when birds are given a 12-h conventional photoperiod prior to a transfer to intermittent lighting after egg production has peaked (24).

It is doubtful whether free-running can be expected when short cycles are used in commercial houses that have some light leakage, and which may also have a marked diurnal temperature cycle.

Although there is a feed saving of 3 to 4%, there is no improvement in feed conversion efficiency and no reduction in end-of-lay body weight.

Reading System

A novel short-cycle regimen was developed at the University of Reading to use the increased egg weight and improved shell quality characteristics of short-cycles to offset the frequently experienced reduction in egg weight of a *Biomittent* programme whilst retaining its benefits of lower feed intake and improved feed conversion (25). The Reading System, which involves giving birds a 0.25L:0.75D cycle continuously, has been reported to result in a 3% reduction in egg numbers, a 2% increase in egg weight, a 3% improvement in shell thickness, and a 6% saving in feed intake compared with a conventional step-up to 15 h (Table 4.2).

An additional benefit of most forms of short-cycle intermittent programmes is that they very effectively control infestations of red mite (26).

Long cycles

Twelve-hour cycles such as 2(2L:10D) were studied to explore the possibility that further reductions in illumination could be made to produce greater electricity savings than those for the asymmetrical Cornell system (27, 28). However, the resulting performance was similar to that for short-cycle, rather than asymmetrical intermittent lighting, with a reduction in egg numbers, egg output and feed intake, but an improvement in egg weight and shell quality (Table 4.2). In contrast to short-cycle programmes, there is an entrainment of egg-laying with most eggs being laid during one of the 10-h scotoperiods (29). This suggests that despite each photoperiod and each scotoperiod being the same size, hens uniformly adopt one scotoperiod as night and perceive the other as being part of a subjective day.

Mortality

Generally, any intermittent lighting programme will have a lower incidence of mortality than one that gives the equivalent solidly illuminated photoperiod (30). Typically, liveability improves by 4% for each 1-h reduction in illumination (Figure 4.4). However, the improvement is not a consequence of intermittent lighting *per se* but of the reduced amount of light. Where there is no reduction in the amount of light, liveability will be similar to conventional controls, for example, liveability for laying hens given a 4(4L:2D) regimen was reported to be similar to 16L:8D controls (8).

General summary

Asymmetrical regimens maintain egg production but save feed and electricity. *Biomittent* lighting has greater feed and electricity savings but with a risk of losing egg weight (especially if introduced too soon after transfer to the laying house). Short-cycles (e.g. 3L:3D) improve egg weight and shell quality but at the expense of egg numbers (if introduced before peak egg production), with the Reading System achieving the largest improvement in feed conversion efficiency.

Intermittent and conventional lighting programmes are interchangeable, so if circumstances change, such as a requirement for egg numbers rather egg size, hens can successfully be transferred back to a normal lighting schedule, with egg output returning to a conventional numbers:weight ratio within 7 d.

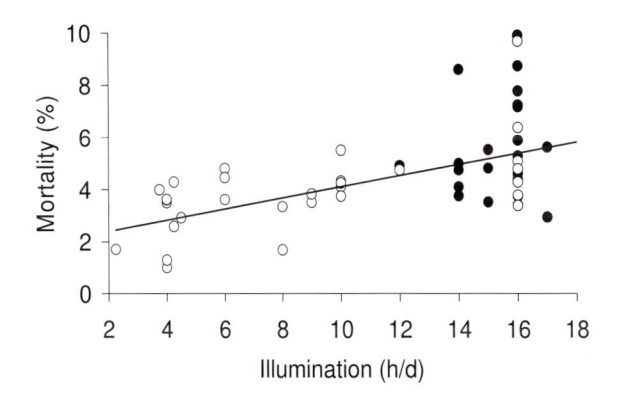

Figure 4.4 *The incidence of mortality during an annual egg-laying cycle for egg-type hybrids given any type of intermittent lighting (○) and for solidly illuminated controls (●). Data from 36 sets adjusted for differences from the mean of 5.4% by least squares analysis (30).*

Broiler breeders

Rearing period

Although broiler breeders are normally reared on a solid 8-h daylength, it has been reported that sexual maturity was delayed by about 3 d

Table 4.3 *Comparative performance of male and female broilers given 7 d of 23-h photoperiods then transferred to an intermittent regimen to 6 or 8 weeks of age (conventionally lighted controls =100). Mean ±SEM (number of comparisons in parentheses) (derived from 36).*

Regimen (h)	Body weight	Feed intake	Feed conversion efficiency
8(1L:2D)	101.3±0.5 (4)	96.7±1.9 (4)	104.9±1.9 (4)
6(1L:3D)	101.5±0.9 (13)	95.5±0.6 (13)	106.4±1.2 (13)
12(0.25L:1.75D)	100.3±0.4 (8)	97.2±0.6 (8)	103.3±1.0 (8)

when they were grown on a *Biomittent*-type 6, 8 or 10-h regimen. In the subjective day, the first hour was fully illuminated and the remainder of the day-time hours segmented into 45 min darkness and 15 min light (31). Subsequent to a transfer to a normal 16-h photoperiod at 20 weeks, egg production for a given number of weeks after reaching 50% lay was similar to conventionally reared controls.

Laying period
The provision of a 2L:3D:10L:9D asymmetrical programme from 24 weeks of age has been shown to result in broiler breeders having similar egg production to, but heavier egg weight and better shell quality than, 14L:10D controls (32). However, the increased egg size and shell quality did not significantly improve hatchability or increase the production of day-old chicks.

Broilers
Generally, it is only short-cycle intermittent programmes that are used in growing broilers, with 8(1L:2D), 6(1L:3D) and 12(0.25L:1.75D) being the most common. These are not usually introduced until the birds have had 7 d of 23-h photoperiods. Feed intake and growth are initially depressed in both males and females when they are transferred to intermittent lighting. Subsequently, higher feed intakes and faster growth result in male body weight being similar to, or heavier than, 23-h photoperiod controls by 6 weeks (1L:3D (33), 1L:2D (34)). The compensatory response is slower in females and, though they have caught up with controls by 8 weeks, intermittently lit females usually have lower body weights than conventional controls at 6 weeks. The interaction between lighting type and sex at 6 weeks is a consequence of higher plasma concentrations of growth hormone occurring in both sexes, but higher testosterone concentrations occurring only in the males (35).

Intermittently lit broilers convert feed more efficiently than 23L:1D birds (Table 4.3, 36). Although this is mainly achieved through a reduction in activity-related energy expenditure, there is also a marginal improvement in the

Table 4.4 *Mean daily energy intake, heat production, and energy retention, and mean hourly heat production during a scotoperiod and photoperiod between 3 and 6 weeks in male broilers exposed to 6(1L:3D) or 23L:1D lighting (37).*

	Lighting regimen	
	23L:1D	1L:3D
Gross energy intake (GE) ($kJ.kg^{0.75}.d$)	2024	2001
Metabolisable energy (ME) ($kJ.kg^{0.75}.d$)	1405	1422
ME/GE	0.695	0.710
Heat production		
Total ($kJ.kg^{0.75}.d$)	848	864
Activity ($kJ.kg^{0.75}.d$)	101	90
Retention		
Energy ($kJ.kg^{0.75}.d$)	557	558
Protein ($kJ.kg^{0.75}.d$)	283	286
Fat ($kJ.kg^{0.75}.d$)	274	272
ME for maintenance ($kJ.kg^{0.75}.d$)	510	525
Heat production: Scotoperiod		
Total ($kJ.kg^{0.75}.h$)	30.6	33.9
Activity ($kJ.kg^{0.75}.h$)	1.6	2.3
Photoperiod		
Total ($kJ.kg^{0.75}.h$)	35.6	42.2
Activity ($kJ.kg^{0.75}.h$)	4.4	8.4

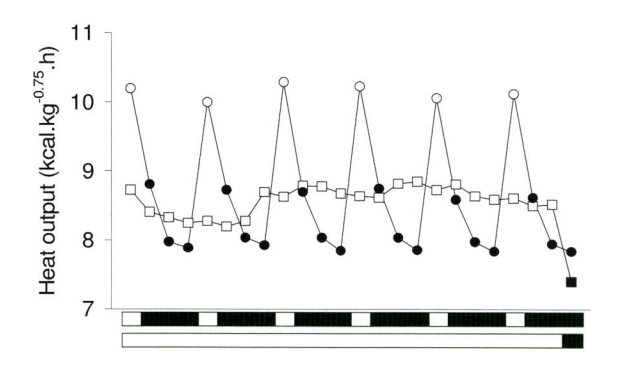

Figure 4.5 *Mean hourly total heat production between 3 and 6 weeks of age for female broilers exposed to 23L:1D (□ = light, ■ = darkness) or 6(1L:3D) lighting (○ = light, ● = darkness), open bars represent light, and solid bars darkness (37).*

metabolisability of the diet (Table 4.4, 37). Mean heat production is higher for intermittently lit broilers than for controls during light and darkness, but the lower production during a scotoperiod than in a photoperiod in both types of lighting and the greater amount of darkness in an intermittent regimen results in similar daily energy expenditures (Table 4.4, 37). However, intermittent lighting results in a very different diurnal pattern of heat output from conventionally lit broilers (Figure 4.5). Reduced oxygen consumption and lower heat production during the dark periods (38) and transitory slower growth following its introduction at about 7 d (34) are mechanisms that might explain the reduction in the incidence of ascites and hydropericardium that occurs when broilers are grown on 8(1L:2D) or 6(1L:3D) lighting cycles compared with a more conventional 23L:1D regimen (39).

Short-cycle programmes have also been reported to improve a broiler's homeostatic regulation of the immune response through the enhancement of splenocyte proliferation, in particular, increased percentages of $CD3^+$, $CD4^+$, and $CD8^+$ subpopulations (39). This is a likely consequence of the extended melatonin release by the pineal gland induced by the larger amount of darkness in intermittent lighting programmes compared with a 23L:1D schedule.

Data for males and females in Table 4.3 show that there is little difference between the beneficial effects of the 8(1L:2D) and 6(1L:3D) regimens on performance, but that the benefits from a 12(0.25L:1.75D) programme are smaller and limited to a reduction in feed intake and more efficient feed conversion, with no increase in body weight (33).

Breeding turkeys

Intermittent lighting programmes can be used for turkey hatching egg production. There needs to be a synchronisation of egg-laying, and so only asymmetrical intermittent programmes are used so that all birds adopt the same light-dark interface as dusk for setting the ovulatory cycle, e.g. 4(1L:2.25D):1L:10D. When intermittent lighting is introduced immediately following the pre-production period of short days, floor-egg-laying and nest training can be problems (41). However, these difficulties are resolved by giving hens a 14L:10D regime for 4 weeks so that they become accustomed to using nests before they are transferred to intermittent lighting. Sexual maturity, egg production, the incidence of broodiness and hatchability traits are not significantly different from 14L:10D controls (42). When the intermittent schedule was modified to accommodate attendant working hours by providing different size scotoperiods during the 14-h subjective day, for example, 1L:4D:4L:1D:1L:2D:1L:10D, reproductive performance was similar to 14L:10D controls, but there was a significant increase in the proportion of eggs laid on the floor (43).

The use of a simpler intermittent schedule, 2L:12D:2L:8D, to photostimulate first year turkeys at 32 weeks and recycled turkey hens at 87 or 102 weeks following 8 weeks of 8L:16D lighting was reported to delay age at 50% lay by 4 d compared with 16-h conventional controls (44). However, over a 20-week production period, the intermittent birds laid 8 more eggs, because of better persistency, and had higher late-season fertility than the controls. These findings are not surprising because the birds would have adopted the 12-h scotoperiod as night and perceived the regimen as a 12-h subjective day (2L:8D:2L:12D not 2L:12D:2L:8D), and a transfer from 8 to 12 h rather than to 16 h would be expected to result in slower sexual development but improved egg production and increased egg weight (Figure 3.30, page 42). The results are similar to those reported for broiler breeders transferred from 8-h to either 12 or 16-h days (see page 37).

A 2L:8D:2L:12D regimen has been used temporarily between 26 and 32 weeks, before a permanent transfer to 16L:8D lighting, to increase early egg size by almost 4 g (46). The hens took 9 d longer to reach 50% lay than birds transferred to a 16-h photoperiod at 26 weeks (due to the transfer to an apparent 12-h day), but matured 16 d earlier than birds transferred to 16 h at 30 weeks. A regression of initial egg weight on age at 50% lay for the three groups indicates that the increase in egg weight was a function of the delay in sexual development (+0.2g/d delay in maturity) and not intermittent lighting itself. The same result could have been expected had the birds been transferred to a 12L:12D schedule at 26 weeks.

The provision of a 2L:9D:2L:11D regimen (equivalent to 13L:11D) for male turkeys from 34 weeks has been shown to result in similar semen volumes and sperm numbers to, but better persistency of normal sperm percentage and lower feed intake than, 15L:9D controls (47). The maintenance of a high percentage of normal sperm may be the male equivalent of the improved end-of-cycle fertility observed in hens given 2L:8D:2L:12D lighting (44). It is possible that both of these improvements in reproductivity at the end of the season were a result of a delayed onset of adult photorefractoriness. In starlings the initiation of sexual development and the processes leading to subsequent gonadal regression are thought to begin simultaneously on the first long-day (48), and so when seasonal breeding birds undergo slower sexual development in response to a mildly stimulatory photoperiod, a slower development of adult photorefractoriness is to be expected (page 16).

Table 4.5 *Comparative performance of male turkeys given ≤ 4 weeks of conventional lighting then transferred to a short-cycle intermittent regimen until 18 to 24 weeks of age (conventional 23-hour controls =100). Mean ±SEM (number of comparisons in parentheses) (50, 51, 52, 54).*

Trait	Comparison
Final body weight	105.4±1.5 (6)
Feed intake	104.8±1.4 (6)
Feed conversion efficiency	100.6±0.8 (6)

Testes weights, semen production, mean plasma testosterone and LH (pulsatile in male turkeys) concentrations, and peak amplitude and length of these hormones under 8(1L:2D) lighting were reported to be similar to 14L:10D, but the number of LH peaks per 12 h sampling period for intermittently illuminated toms (6.8±1.4) was significantly fewer than for conventionally lit controls (11.7±0.6) (49).

It can be concluded that breeding turkeys respond to asymmetrical intermittent lighting in a comparable way to domestic fowl, with reproductive performance being similar to that expected for birds given the solidly illuminated equivalent of the subjective day. (Table 4.2).

Growing turkeys

Turkeys given 8(1L:2D), 6(1L:3D), 6(2L:2D), 4(2L:4D) or 4(4L:2D) schedules following an initial 2 to 4 weeks of conventional lighting have higher feed intakes and heavier body weights than conventionally illuminated controls, irrespective of the light:dark cycle (Table 4.5). The benefits are normally greater for males than females because the intermittent schedules are sexually stimulatory and the older killing ages for males allow the birds time to respond to the photostimulated rise in testosterone and the concomitant increase in feed intake. Although there is normally a small improvement in feed conversion efficiency, most of the extra growth emanates from the increase in feed intake. In contrast to broilers (page 62), there is minimal information on the effects that short-cycle programmes have on energy expenditure and endocrine rhythms in turkeys, and so the contributions that modifications of these may make towards the increased growth is unknown.

Asymmetrical intermittent programmes also increase body weight compared with 23L:1D controls and, again, this is primarily achieved through an increase in feed intake (50). A *Biomittent*-type regimen in which a 12-h day was segmented into 1L:0.5D resulted in precocious sexual development in males between 16 and 20 weeks and produced heavier body weights than controls held on non-stimulatory 8-h photoperiods. However, the extra weight gain was produced through an improvement in feed conversion efficiency and not by an increase in feed intake (53).

Although liveability and leg disorders are usually unaffected by intermittent lighting, there is one report in which a 2L:4D repeating schedule given from 1 d reduced the incidence of leg anomalies compared with a lighting programme stepping down 1h/d to 15 h (54).

• Ahemeral programmes •

Ahemeral regimens are used only for egg-laying birds because their main function is to modify the ovulatory cycle, so that egg output can be repackaged to meet market requirements. Short-cycle regimens could be regarded as ahemeral, in the sense that they are light-dark cycles which do not entrain the bird to 24 h, but responses to the two types of lighting are very different. Laying hens on short-cycle inter-mittent programmes are unable to form a day and night, and so they free-run and the flock lays eggs at random across the 24-h day; and this phenomenon still occurs with 8L:8D cycles (55). In longer ahemeral regimens, the hen is able to recognise a day and night and, with cycles between 21 and 30 h, there is complete entrainment of egg-laying (56). With 36-h cycles the hen reads the two cycles as making up three 24-h days: it treats 18L:18D:18L:18D as though it were 18L:54D. This section deals only with programmes where the cycle length is such as to cause entrainment of egg-laying.

The photoperiodic effect of ahemeral cycles
A surprising feature of those cycles longer than 24 h that entrain the internal biological clock of the bird is that they are more photostimulatory than a 24-h cycle containing the same photo-period. For example, a bird transferred from 10L:14D to 10L:18D responds as though it had been moved to a 14-h day (14L:10D). The reason for this is that the biological clock does not 'run slow' under the 28-h cycle but resets itself following each dusk signal. After 18h of darkness, the bird is expecting 6 h of light before the next dusk, which would be a non-stimulatory schedule. However, the lights stay on for a further 4 h after the expected dusk, illuminating the photoinducible phase and producing a positive gonadotrophin response. The effective photoperiod in an ahemeral cycle of between 25 and 30 h is (P + C - 24), where P is the actual photoperiod and C is the cycle length (h) (58). This rule is of greater importance when returning a flock from an ahemeral cycle to a normal 24-h day. For example, if layers have been kept for a time on 14L:13D, perhaps to increase early egg size (see below), and they are to be returned to 24-h cycles to limit further increases in egg size, their new regimen must be 17L:7D to avoid a reduction in the effective photoperiod. If the photoperiod is not increased, and the birds are returned to a 14L:10D schedule, they will respond as if they had been given a 3-h decrease in photoperiod.

Growing pullets
Circadian rhythms have been reported to adapt to light-dark cycles of between 21 and 30 h, and so it might be expected that growing pullets reared on non-24-h schedules will grow and reach sexual maturity differently from pullets illuminated conventionally (57). However, when pullets were given a 14-h photoperiod within a 21, 24 or 28-h cycle there were no significant differences in growth profile, 16-week body weight or age at first egg, even though the three groups would have had 160, 140 and 120 subjective days, respectively, to 20 weeks of age.

Immature pullets transferred from short days at 17 or 22 weeks to 27 or 28-h light-dark cycles containing an effective photoperiod equivalent to 24-h controls (e.g. 9L:18D compared with 12L:12D) matured at a similar age to conventionally illuminated controls (58, 59). However, a transfer to a 27-h cycle with no change in photoperiod (6L:18D to 6L:21D) advanced maturity because it was treated as a 3-h increment in effective photoperiod (58).

Laying hens
Cycles of less than 24 h are really only of interest to geneticists, who may use them when selecting for egg numbers, but they are of no value to the commercial egg-laying industry. Laying hens can only have one pre-ovulatory LH surge in each light-dark cycle, and so hens that have the ability to lay eggs at intervals of less than 24 h are prevented from doing so by the constraints of the 24-h solar day (60). If such individuals are

Table 4.6 *Theoretical maximum rates of lay (% per 24 h) for hens with different natural rhythms of ovulation kept under various light-dark cycles (61). Bold data indicate where production is limited by the light-dark cycle being longer than the natural rhythm.*

Natural rhythm (h)	Light-dark cycle (h)					
	22	23	24	25	26	28
22	109	**104**	**100**	**96**	**92**	**86**
24	91	94	100	**96**	**92**	**86**
26	82	82	83	87	92	**86**
28	76	76	75	76	77	86

transferred to a light-dark cycle that matches their own natural rhythm, rate of lay should be maximised. However, egg production will fall for other birds in the flock that have a natural rhythm longer than the ahemeral cycle (Table 4.6, 61). The effects on mean rate of lay for a flock of hens will therefore depend on the proportion of individuals whose natural rhythm is close to the chosen light-dark cycle.

In contrast, cycles greater than 24 h are of practical value to commercial poultry keepers because they can be used to increase egg weight when there is a demand for larger eggs or to ameliorate a shell quality problem by increasing shell thickness. However, welfare regulations may prohibit the use of ahemeral schedules in some countries; for example, a lighting regime must follow a 24-h rhythm in the EU (62).

Egg production

The first reported use of ahemeral lighting for laying hens appears to have been in 1941 (63). It was concluded that a 14L:12D schedule improved egg production quite dramatically, compared with 14L:10D controls, by lengthening clutches and synchronising the light-dark cycle with the hen's natural (26-h) ovulation cycle. Although this is supported by the calculated maximum rates of lay in Table 4.6, where the maximum egg production for birds with a 26-h natural rhythm is predicted to rise from 83 to 92% when they are transferred to a 26-h ahemeral cycle, the subsequent elucidation of the way in which the hen interprets photoperiods within an ahemeral cycle (58) questions these conclusions. A 14L:12D cycle is photoperiodically equivalent to 16:8D, and the improvement in egg production reported in 1941 was almost certainly a response to the 2-h increment in the effective photoperiod. The failure to match the effective photoperiod under ahemeral lighting with the actual photoperiod used in the conventional regimen in other early investigations into ahemeral lighting may also explain some of the anomalies reported in the literature. However, subsequent selection for egg numbers has produced hens that are now able to ovulate at intervals very close to 24 h, and so transferring modern hybrids to a 26-h cycle early in lay will result in a drop in peak egg production from nearly 100 to just under 92% (Table 4.6).

The average interval between ovipositions within a sequence increases progressively by about 15 min per month between 28 and 56 weeks (64). It follows, therefore, that the adverse effect on rate of lay following a transfer to ahemeral lighting reduces as the hen ages (65), to the point where there is either no effect or even an improvement when the interval between eggs within a sequence equals the length of the ahemeral cycle (66).

In addition to being dependent upon the age at which birds are transferred to ahemeral lighting, the effects on egg production vary with the length of the ahemeral cycle (67). Figure 4.6 shows that egg production decreases by 1.7% of conventional controls when hens are changed to an ahemeral regimen before 35 weeks of age, but that it consistently improves when the change is made after 55 weeks. The application

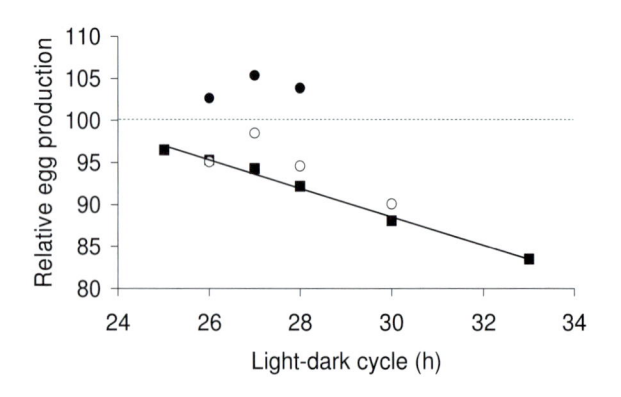

Figure 4.6 *Relative egg production (conventional lighting = 100) for hens changed to different light-dark cycles at < 35 weeks (■), between 35 and 55 weeks (○), or after 55 weeks (●). A linear regression is fitted to the < 35-week changes (67).*

of ahemeral lighting between 35 and 55 weeks has variable effects on egg production because at this stage some birds have a potential interval between ovipositions that is shorter than the applied light-dark cycle, whilst others have a natural rhythm that is longer.

With the exception of a change at 22 weeks (when the natural interval between ovipositions is likely to have been less than 24 h for some birds), transferring hens to a 21 or 23-h light-dark cycle invariably results in a reduction in rate of lay because the light-dark cycle is shorter than the interval between ovipositions (Table 4.6) (67).

Egg weight and shell thickness

Deposition of yolk material from the liver into the ovary occurs at a steady rate and is independent of the time that the previous egg spends in the oviduct or when it is laid (68). Thus, the longer interval between ovipositions for hens on ahemeral lighting results in the production of larger yolks, followed by proportionately more albumen and, ultimately, heavier eggs. An improvement in shell quality also occurs because, although the egg spends longer in each part of the oviduct, it is held for almost an hour longer in the shell gland; 20.7 h under 27-h cycles compared with 19.8 h for 14L:10D controls (69).

The improvements in shell quality and egg weight are causally linked to oviposition intervals, but not necessarily to a reduction in rate of lay. There is an apparent association when birds are given ahemeral lighting early the lay, but improvements in shell thickness still occurred when a transfer from 16L:8D to 12L:16D cycles was made at 60 weeks of age although rate of lay actually increased in those hens with a sequence length of less than 6 eggs (66). Additionally, the mechanisms for the increases in egg weight and shell thickness (mg/cm^2) are not the same because, following a change from 16L:8D to 12L:16D at 30 weeks of age, increases in shell thickness were seen after 2 d, but increases in egg weight did not occur until 4 d after the light change. Allometric increases in shell thickness for three genotypes were between 10 and 15% greater than the increases in egg weight (70). This report also noted that

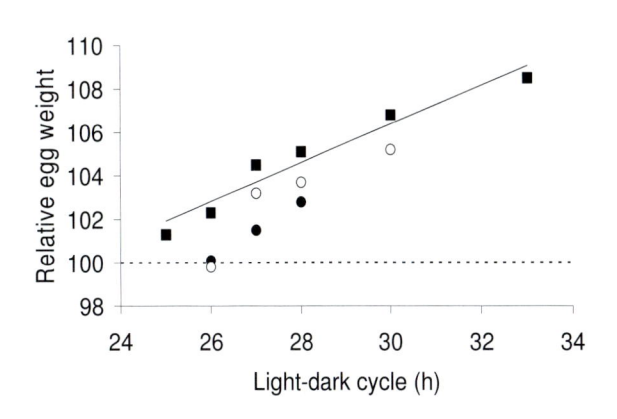

Figure 4.7 *Relative egg weight (conventional lighting = 100) for hens changed to different light-dark cycles at younger ages than 35 weeks (■), between 35 and 55 weeks (○), or after 55 weeks (●). The regression line is fitted to all data, except those for hens transferred to 26-h cycles after 35 weeks. Differences between ages adjusted by least squares to < 35-week changes. (67).*

whereas the increases in egg weight were similar for the three genotypes (4.2, 4.9, and 5.9 g), improvements in shell thickness index varied significantly (0.4, 4.7, and 12.1 mg/cm^2).

As might be expected, the increase in egg weight depends on the length of the ahemeral cycle as well as the age at which it is introduced. Figure 4.7 shows that, with the exception of transfers to a 26-h cycle at > 35 weeks, egg weight increases by about 1% of conventional

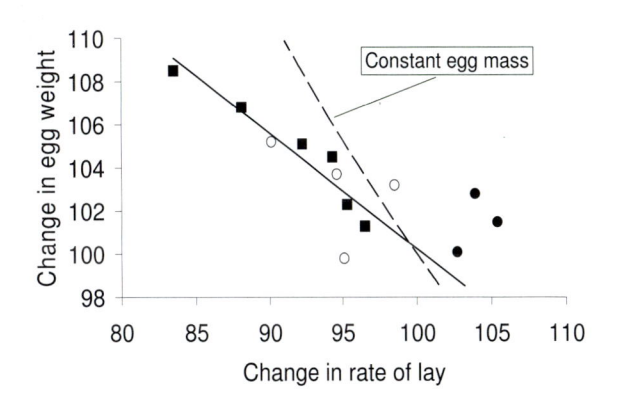

Figure 4.8 *A regression of change in egg weight on change in rate of lay (conventional lighting = 100) for hens exposed to different light-dark cycles at < 35 weeks (■), between 35 and 55 weeks (○), or after 55 weeks (●). The solid regression line is for changes made < 35-week, and the broken curve represents a constant egg mass output (67).*

controls for each 1-h extension of the light-dark cycle, irrespective of age of application, but that ahemeral lighting results in smaller increases in egg weight when it is applied at older ages (67).

Egg weight is consistently improved by about 1 to 2% by a change to 23-h cycles, which, for most hens, causes shorter clutches and longer intra-clutch intervals (67).

Egg mass output

When ahemeral lighting is applied below 35 weeks of age, egg mass output decreases because relative increases in egg weight are only half the corresponding relative decreases in rate of lay (Figure 4.8). Interestingly, the regression for the < 35-week hens passes naturally through the 100/100 point. However, when introduced after 55 weeks, positive responses in rate of lay and egg weight result in an increase in egg output. Between 35 and 55 weeks, the results are variable, due presumably to some hens having a natural rhythm shorter, and some longer, than the light-dark cycle (67).

Oviposition time

Melatonin rhythms are strongly correlated with the timing of the open-period for the pre-ovulatory LH release and, as a consequence, the time of egg-laying (page 30). Although any light suppresses melatonin release, and synthesis of melatonin does not occur in 'unexpected' darkness, melatonin cycles are entrained to an ahemeral lighting regimen within two cycles, resulting in a shift in the timing of oviposition two cycles later (71). In conventional 24-h cycles, the ovulatory cycle is entrained by as little as 15-min of light or by 5 h of darkness. However, hens on ahemeral cycles require stronger cues to synchronise egg-laying and, under 21-h cycles, the minimum photoperiod to achieve full phase setting is 3 h and the minimum scotoperiod is 9 h (56). Under 30-h cycles, the minimum light and dark periods necessary for entrainment are 8 and 12 h, respectively. If the appropriate minima are provided, there is complete entrainment to non-24-h light-dark cycles, though the egg-laying time will, as in 24-h schedules, be dependent upon the light-dark cycle length. Mean oviposition time (MOT) can be predicted using the following equations (72):

For < 24-h cycles
$$MOT = -4.97 + 0.740C + 4.482S - 0.175CS$$

For > 24-h cycles
$$MOT = 64.62 - 2.161C + 0.268S$$

where MOT = mean time of lay, C = cycle length (h), and S = scotoperiod (h).

Feeding activity and feed intake

Feed activity in hens exposed to 21, 24 or 30-h light-dark cycles that contain either a 12-h photoperiod or a 10-h scotoperiod is controlled by a circadian clock that is set by the preceding light-dark interface (73). Peak activity occurs at the same time relative to the previous dusk, irrespective of the light-dark cycle, and is not related to the ovulatory cycle that is modified by the ahemeral regimen.

Notwithstanding that in two studies of 28-h cycles, ahemerally lit birds had a 2-3 g significantly lower intake than conventional controls (74), a paired t test of 8 sets of data for ahemeral cycles of between 26 and 30 h shows that feed intake is generally not significantly different from that under 24-h regimens (74-79).

Practical application

Ahemeral lighting is now seldom used because larger egg size and improved shell quality can be achieved more conveniently by using a short-cycle intermittent programme. It also contravenes European Union welfare regulations. However, when it was used, it was almost always a 28-h cycle because six 28-h cycles exactly fit a 7-d solar week. The main disadvantage of ahemeral lighting was its asynchrony with the 7-d working week; by the middle of the week the hen's day was the worker's night and the worker's day, the hen's night. It was possible to partially overcome this problem by the use of bright-dim rather than light-dark cycles. The hen interprets the bright period as day and the dim period as night, provided there is sufficient contrast between the intensities (80). However, the ratio of illuminance necessary to achieve full entrainment of egg-laying depends on the ahemeral cycle; the further it is below 24 h or above 27 h, the greater the contrast required for synchronising the ovulatory cycle (Figure 4.9, 80). Whereas a 10:1 bright:dim ratio will entrain

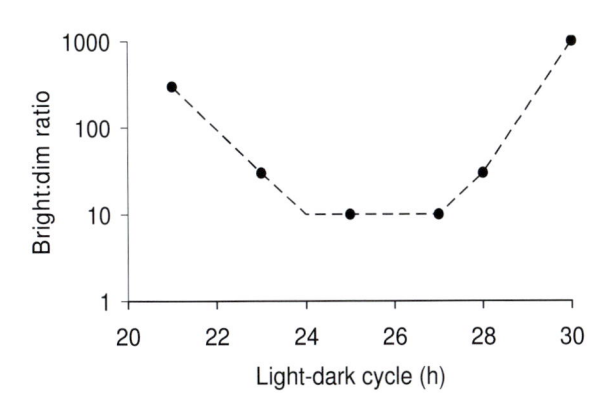

Figure 4.9 *Bright:dim ratios required to synchronise egg-laying in hens exposed to various ahemeral light-dark cycles, each incorporating a 14-h photoperiod (80).*

egg-laying for hens on cycles between 24 and 27 h, a 30:1 contrast is needed for entrainment to a 28-h cycle, a 300:1 for a 21-h, and 1000:1 for a 30-h cycle. Entrainment has also been achieved using cycles of white (day) and blue (night) light (65).

Fortunately, conventional and ahemeral lighting programmes are interchangeable, provided the appropriate adjustment is made to the photoperiod (see page 65). This presents the opportunity to use an ahemeral regimen at the beginning of the laying period to increase egg weight (74, 81), to revert to a conventional 24-h cycle when egg weight is normally satisfactory, and to re-introduce ahemeral lighting if required later in the laying year to improve shell quality (81, 82).

Broiler breeders

The transfer of broiler breeders to 14L:13D at 24 weeks from a step-down lighting programme which reached 14L:10D at 24 weeks resulted in similar sexual maturity, 1.5 g heavier eggs at 32 weeks and improved shell thickness, but yielded 24 fewer total eggs and 21 fewer hatching eggs (classified by weight) to 64 weeks of age compared with conventional 14L:10 birds (32). Surprisingly, the improved shell quality had no effect on hatchability. Although the experimental birds would have interpreted 14L:13D as 17L:7D, so confounding the extended light-dark cycle with an increased photoperiod, the reduction in egg numbers, increase in egg weight and production of thicker shells are similar to the responses of egg-type pullets to ahemeral lighting introduced before or soon after sexual maturity. However, a major disadvantage for the ahemeral treatment was that 50% of the eggs were laid on the floor and so unsuitable for use as hatching eggs.

In separate experiments in which broiler breeders were transferred to 28-h cycles at 22 or 32 weeks of age to improve initial egg weight, there were also depressions in rate of lay, even though the photoperiod had been appropriately reduced in one of these trials (84, 85). When the introduction of ahemeral lighting was delayed until 44 weeks of age, rate of lay increased (84). This agrees with the response of egg-type hybrids transferred to ahemeral cycles when their natural cycle is longer than the light-dark cycle (66). In contrast, rate of lay was reported to have been lower than controls in another experiment when broiler breeders were changed from 15L:9D to 15L:13D at 47 weeks of age, though egg weight was predictably increased (86). Although this may suggest that broiler breeders sometimes respond differently from egg-type hybrids when ahemeral lighting is applied late in the laying cycle, these birds would have interpreted the change as one from a 15 to

Table 4.7 *Relative age at sexual maturity, egg numbers to 53 weeks, floor eggs, mean egg weight, and shell thickness for turkey breeders transferred from a conventional 8-h photoperiod to 21, 23, 26, 27, 28 or 30-h light-dark cycles containing a 15-h photoperiod at 30 weeks (24-h controls = 100) (87).*

Light-dark cycle (h)	Age at 50% lay	Egg numbers	Floor eggs	Mean egg weight	Shell thickness
21	123	83	220	104.3	101.3
23	105	108	73	100.6	98.3
26	104	102	134	100.6	99.9
27	95	99	169	101.5	101.7
28	112	96	317	101.3	102.6
30	126	79	341	104.6	105.9

a 19-h photoperiod, and it is possible that the perceived increment advanced the onset of photorefractoriness and depressed egg production. This indicates that provision of the correct photoperiod, when a change is made from conventional to ahemeral cycles, is more important for broiler breeders than for egg-type hybrids.

Breeding turkeys

Transferring turkeys from a conventional 8L:16D regimen at 30 weeks of age to various cycles between 21 and 30 h has been shown to result in responses similar to those of domestic fowl (87). Notwithstanding that a common 15-h photoperiod was provided in all treatment-groups and would have been perceived differently within each, it is evident that the more extreme the light-dark cycle, the later the sexual maturity, the poorer the egg production to a given age, the larger the eggs, the thicker the shells and the greater the proportion of eggs laid on the floor (Table 4.7, 87).

The significant differences in age at 50% lay were probably as much a consequence of differences in perceived increment in photo-period at photostimulation as to the light-dark cycle itself, and are unlikely to indicate the response to ahemeral lighting *per se*, but will have had a marked influence on the number of eggs produced to a fixed age.

This report concluded that the practical limits for maximising egg weight and shell thickness, though at the expense of egg numbers, are transfers to light-dark cycles of between 23 and 28 h (87). However, different results might have been obtained had the photoperiods been adjusted to give a common perception of a 15-h photoperiod for all treatments.

Table 4.8 *Egg production to 54 weeks for turkeys transferred from 8L:16D to 15L:13D at 26 weeks for 2 weeks, 5 weeks or permanently, or to 15L:9D at 26 or 30 weeks (88).*

Lighting treatment (h)	Total eggs
15L:13D from 26-28 weeks	116.7
15L:13D from 26 to 31 weeks	118.3
15L:13D from 26 to 54 weeks	94.9
15L:9D from 26 weeks	111.3
15L:9D from 30 weeks	102.7

When 15L:13D ahemeral lighting was introduced at 26 weeks of age, the initial egg weight was reported to be superior to that of turkeys transferred to 15L:9D conventional lighting and similar to that of birds conventionally photostimulated at 30 weeks. Although the temporary application of 28-h ahemeral lighting for up to 5 weeks from 26 weeks benefited early egg weight without adversely affecting egg numbers (Table 4.8), a permanent transfer to ahemeral lighting from 26 weeks resulted in a significantly lower rate of egg production to 54 weeks with no beneficial effect on egg weight during the latter stages of the production cycle (88, 89).

• Continuous illumination •

Under conditions of constant temperature and illumination, fowl are released from the confines of a 24-h solar day and show a free-running rhythm of about 25.3 h for many circadian oscillators, including deep body temperature, feeding activity and ovulation (11). When birds are exposed to continuous illumination (LL) at high light intensities (200 lux) or for prolonged periods (> 4 weeks), the circadian rhythm breaks down into an ultradian rhythm of about 5 h (90). However, the responses to LL are modified when other environmental cues such as temperature or feed delivery times are neither constant nor random (91).

Broilers

Most of the data for the effects of continuous light on growth are for early meat-type chickens. Tremendous advances have been made in the growth potential of broiler genotypes since this early work, so the response of modern broilers may be different. Although many reports have shown that the growth of meat-type chickens is maximised if the birds are exposed to LL conditions (e.g. 92), it has become commercial practice to give a 1-h scotoperiod to accustom birds to darkness so that the effects of panic in the event of a power failure are minimised. Thus, continuous illumination is unlikely to be used in modern broiler and turkey growing operations even if welfare regulations allow it.

It has been suggested that if the function of LL was simply to maximise feeding opportunity, then the performance under variable illuminance would be similar to that for constant illuminance (92). However, providing 8 or 16 h of bright light (8-52 lux) and 16 or 8 h of dim light (0.7-1.6 lux) resulted in significantly lower feed intake and lighter body weights than, but similar feed conversion efficiencies to, constant bright or dim light groups. These findings indicate that factors other than feeding opportunity are involved in the maximisation of growth under LL, and that controlled environment housing needs to have good light control to maximise the benefits of continuous lighting. The results are supported by earlier observations that growth under a combination of artificial and natural light is inferior to that in controlled environment facilities (93). Nevertheless, feeding opportunity does play a major role in maximising growth under LL because body weight was lower and feed conversion efficiency inferior when access to feed was limited to 8 h per day (94).

Feeding activity still shows a diurnal pattern in broilers kept in constant LL conditions (95), but locomotor activity is uniformly distributed over the 24-h period (96).

Growing turkeys

There are no recent data for the effect of continuous illumination on turkeys grown to normal slaughter weights but, to 8 weeks, providing LL conditions had no significant effect on body weight, feed conversion efficiency or liveability compared with birds given a 23L:1D or 12L:12D regimen (97). The effects of LL to older ages, in particular males kept to 20 weeks and over, are speculative, but increased growth and improved feed conversion efficiency, compared with turkeys kept on short days, might be expected because of the stimulatory effect of LL on sexual development.

Cockerels and laying hens
Reproduction
Continuous illumination has minimal effect on rate of egg production or fertility in hens (98, 99) but a marked adverse effect on fertility in

Table 4.9 *Rate of lay and fertility for domestic fowl maintained on continuous illumination or 14-h photoperiods from hatch, and mated naturally or artificially (99).*

Photoperiod (h) for hens (F) and cockerels (M)	Rate of lay (%)	Fertility (%)
Natural mating		
F24-M24	54.5	34.0
F24-M14	58.7	73.0
F14-M24	60.2	20.2
F14-M14	60.5	72.2
Artificial mating		
F24-M24	68.0	92.2
F24-M14	69.0	95.1
F14-M24	70.0	96.7
F14-M14	63.0	94.2

males (99). However, the poor fertility in males is probably the result of some behavioural problem, possibly impaired vision (see pages 104-105) or lethargy, because artificial insemination of semen from the same cockerels produced normal fertility (Table 4.9).

Feeding activity and feed intake
Feed intake shows a mild circadian rhythm under LL conditions, but the rhythm is more pronounced in mature than in immature pullets because of the strong influence that the ovulatory cycle has on feeding activity (100, 101). Servicing of the birds can also have an effect (91).

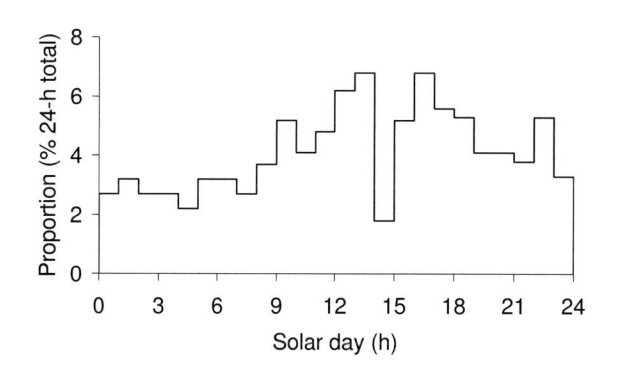

Figure 4.10 *Hourly distribution of egg-laying for continuously illuminated hens (104).*

Oviposition

In the absence of a light-dark interface to act as a zeitgeber for setting the open-period for pre-ovulatory luteinizing hormone release when hens are lit continuously, ovulation can occur at any time because laying hens hormonally free-run according to an internal biological clock that has a rhythm of about 25.3 h (11). However, despite light providing the strongest signal for synchronising circadian rhythms, other environmental stimuli such as noise, temperature and servicing, can influence birds under LL conditions, and so they continue to lay eggs in clutches, with the first egg of each clutch being laid at a mean time of 12.30 h (102). As a consequence, there is still a diurnal pattern of oviposition in LL, with 46% of eggs laid in the modal 8 h, even when continuous 'white' noise is provided, and a mean oviposition time early in the afternoon (103, 104) (Figure 4.10).

• **Continuous darkness** •

Broilers

Body weights at 4 and 6 weeks of age for male and female broilers transferred to continuous darkness at 7 d were reported to be lower than for birds maintained on 23L:1D lighting because of a temporary depression in feed intake when the birds were first put into the dark (105). However, after the birds had learnt to eat in darkness, compensatory growth resulted in body weights being similar at 8 and 10 weeks (Table 4.10). Feed conversion efficiency and carcass quality were also not significantly different from illuminated controls at 8 and 10 weeks.

Growing pullets and laying hens

Sexual maturity

Growing pullets kept in almost total darkness (DD) from 7 d reached 10% egg production 14±3.4 d later than pullets reared on 6-h photoperiods (106, 107). As a comparison, pullets maintained on 6-h photoperiods matured 3 to 8 d later 12-h birds (106). The DD birds, which were reported to be more docile, had similar mortality, but lower feed intakes that resulted in smaller body weights at 22 weeks and at 10% production (107).

Egg production

Pullets kept in continuous darkness during both the rearing and the laying periods were reported to have inferior rates of lay (59%) to controls reared on 6L:18D lighting and given a step-up regimen from 22 or 25 weeks (73%). Production for pullets reared in darkness but given a step-up lighting programme in lay was intermediate (66%) (107). Pullets exposed to DD conditions throughout their life had significantly higher mortality during lay, but laid larger eggs with thicker shells than birds reared on short days, the usual responses of free-running individuals.

When conventionally illuminated birds were transferred to DD conditions at 38 weeks of age, at a time when mean rate of lay was 60% and some birds had already gone out of lay, others paused lay for about 5 d, whilst some of the non-layers resumed production (108). In a later report, rate of lay for a higher producing genotype transferred from 6L:18D to DD fell from 81 to 64% over a 4-week period, but this contrasted with a decrease from 93 to 58% for other birds transferred from 14L:10D to 6L:18D (109).

These findings show that whilst light is not a prerequisite for sexual development or egg production, total darkness does modify the timing of onset of lay and may result in lower mean rates of lay to a fixed age.

Table 4.10 *Mean feed intake and body weight for male and female broilers maintained on 23L:1D lighting or transferred to continuous darkness (DD) at 7 d (105).*

Age (weeks)	23L:1D	DD from 7 d
Feed intake (kg)		
2	0.30	0.27
4	0.86	0.83
6	1.80	1.76
8	2.92	2.90
10	4.30	4.30
Body weight (kg)		
2	0.19	0.19
4	0.53	0.51
6	1.01	0.96
8	1.53	1.55
10	2.03	2.04

Oviposition time

As with continuous illumination (Figure 4.10), eggs are laid throughout the solar day when hens are exposed to DD conditions (Figure 4.11). Nevertheless, there is still a clear diurnal pattern with 50% of ovipositions in the modal 8 h (random egg-laying would give 33%). However, in contrast to LL hens, DD hens laid most eggs before mid-day, with the average first egg in a sequence being laid at 01.00h and a mean oviposition time of 05.00 h (104). This again indicates the influence of another environmental signal, but either one that is different from that influencing egg-laying in LL hens or one that interacts differently with darkness than with light.

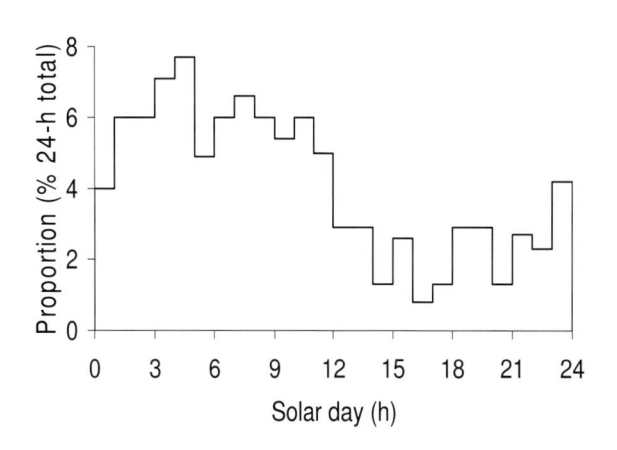

Figure 4.11 *Hourly distribution of egg-laying for hens kept in continuous darkness (104).*

• References •

1. Dobie, J.B., Carver, J.S. & Roberts, J. (1946) Poultry lighting for egg production. *Washington Agricultural Experiment Station, Bulletin no. 471.* pp.1-27.
2. Wilson, W.O. & Abplanalp, H. (1956). Intermittent light stimuli in egg production of chickens. *Poultry Science* 35, 532-538.
3. van Tienhoven, A. & Ostrander, C.E. (1973) The effect of interruption of the dark period at different intervals on egg production and shell breaking strength. *Poultry Science* 52, 998-1001.
4. Rowland, K.W. (1985) Intermittent lighting for laying hens: a review. *World's Poultry Science Journal* 41, 5-19.
5. Lewis, P.D. & Perry, G.C. (1990) Response of laying hens to asymmetrical interrupted lighting regimens: physiological aspects. *British Poultry Science* 31, 45-52.
6. Lewis, P.D., Perry, G.C. & Tuddenham, A. (1987) Noise output of hens subjected to interrupted lighting regimens. *British Poultry Science* 28, 535-540.
7. Lewis, P.D., Perry, G.C. & Cockrem, J.F. (1989) Plasma concentrations of melatonin in layers under interrupted lighting. *British Poultry Science* 30, 968-969.
8. Torges, von H-G., Rauch, H-W. & Wegner, R-M. (1981) Intermittierende beleuchtung von Legehennen und ihr einfluß auf legeleistung, eiqualität, eiblage- und futteraufnahmerhythmik. *Archiv für Geflügelkunde* 45, 76-82.
9. Snetsinger, D.C., Engster, H.M. & Miller, E.R. (1979) Intermittent lighting for laying hens. *Poultry Science* 58, 1109.
10. Mongin, P., Jastrzebski, M. & van Tienhoven, A. (1978) Temporal patterns of ovulation, oviposition and feeding of laying hens under skeleton photoperiods. *British Poultry Science* 19, 747-753.
11. Kadono, H., Besch, E.L. & Usami, E. (1981) Body temperature, oviposition, and food intake in the hen during continuous light. *Journal of Applied Physiology* 51, 1145-1149.
12. Fox, S. & Morris, T.R. (1958) Flash lighting for egg production. *Nature* 182, 1752-1753.

13. Sauveur, B. & Mongin, P. (1983) Performance of layers reared and/or kept under different 6-hour light-dark cycles. *British Poultry Science* 24, 405-416.
14. Lewis, P.D. (1987) Responses of laying hens to interrupted lighting regimes. *PhD thesis*, University of Bristol.
15. van Tienhoven, A., Ostrander, C.E. & Gehle, M. (1984) Response of different commercial strains of laying hens to short total photoperiods in interrupted night experiments during days of 24 and 28 hours. *Poultry Science* 63, 2318-2330.
16. Berman, A. & Meltzer, A. (1978) Metabolic rate: its circadian rhythmicity in the female domestic fowl. *Journal of Physiology, London* 282, 419-427.
17. Harrison, P.C. & Odom, T.W. (1980) Researchers report workable 5-on, 19-off lighting program. *Feedstuffs* 9 June, pp. 14-15.
18. Lewis, P.D. & Perry, G.C. (1986) Effects of interrupted lighting regimens on the feeding activity of the laying hen. *British Poultry Science* 27, 661-669.
19. Morris, T.R., Midgley, M. & Butler, E.A. (1990) Effect of age at starting Biomittent lighting on performance of laying hens. *British Poultry Science* 31, 447-455.
20. Leeson, S., Walker, J.P. & Summers, J.D. (1982) Performance of laying hens subjected to intermittent lighting initiated at 24 weeks of age. *Poultry Science* 61, 567-568.
21. Midgley, M., Morris, T.R. & Butler, E.A. (1988) Experiments with the Bio-Mittent lighting system for laying hens. *British Poultry Science* 29, 333-342.
22. North of Scotland College of Agriculture (1985) Development trials on lighting regimes for layers. *Technical Note 87*, pp. 1-3.
23. Morris, T.R., Midgley, M, & Butler, E.A. (1988) Experiments with the Cornell intermittent lighting system for laying hens. *British Poultry Science* 29, 325-332.
24. Belyavin, C.G. (1986) Changing the light pattern can increase margins. *Poultry Misset* 52, 982-991.
25. Morris, T.R. & Butler, E.A. (1995) New intermittent lighting programme (the Reading System) for laying hens. *British Poultry Science* 36, 531-535.
26. Stafford, K.A., Lewis, P.D. & Coles, G.C. (2006) A preliminary investigation into the role of intermittent lighting for the control of red mite, *Dermanyssus gallinae*, infestations in poultry houses. *Veterinary Record* in press.
27. Simons, P.C.M. & Zegwaard, A. (1983) Verlichting in verband met produktiviteit en energiebesparing van leghennen. *Bedrijfsontwikkeling* 14 October, pp. 785-789.
28. van Tienhoven, A. & Ostrander, C.E. (1976) Short total photoperiods and egg production of White Leghorns. *Poultry Science* 55, 1361-1364.
29. Mongin, P. (1980) Food intake and oviposition by domestic fowl under symmetric skeleton photoperiods. *British Poultry Science* 21, 389-394.
30. Lewis, P.D., Morris, T.R. & Perry, G.C. (1996) Lighting and mortality rates in domestic fowl. *British Poultry Science* 37, 295-300.
31. Lewis, P.D. & Gous, R.M. (2006) Various photoperiods and *Biomittent*™ lighting during rearing for broiler breeders subsequently transferred to open-sided housing at 20 weeks. *British Poultry Science* in press.
32. Proudfoot, F.G. (1980) The effects of dietary protein levels, ahemeral light and dark cycles, and intermittent photoperiods on the performance of chicken broiler parent genotypes. *Poultry Science* 59, 1258-1267.
33. Buyse, J., Simons, P.C.M., Boshouwers, F.M.G. & Decuypere, E. (1996) Effect of intermittent lighting, light intensity and source on the performance and welfare of broilers. *World's Poultry Science Journal* 52, 121-130.
34. Buyse, J., Kühn, E.R. & Decuypere, E. (1996) The use of intermittent lighting in broiler raising. 1. Effect on broiler performance and efficiency of nitrogen retention. *Poultry Science* 75, 589-594.
35. Ohtani, S. & Leeson, S. (2000) The effect of intermittent lighting on metabolizable energy intake and heat production of male broilers. *Poultry Science* 79, 167-171.

36. Kühn, E.R., Darras, V.M., Gysemans, C., Decuypere, E., Berghman, L.R. & Buyse, J. (1996) The use of intermittent lighting in broiler raising. 2. Effects on the somatotrophic and thyroid axes and on plasma testosterone levels. *Poultry Science* 75, 595-600.
37. Apeldoorn, E.J., Schrama, J.W., Mashaly, M.M. & Parmintier, H.K. (1999) Effect of melatonin and lighting schedule on energy metabolism in broiler chickens. *Poultry Science* 78, 223-229.
38. Buyse, J., Decuypere, E. & Michels, H. (1994) Intermittent lighting and broiler production. 2. Effect on energy and on nitrogen metabolism. *Archiv für Geflügelkunde* 58, 78-83.
39. Buy, N., Buyse, J., Hassanzadeh-Ladmakhi, M. & Decuypere, E. (1998) Intermittent lighting reduces the incidence of ascites in broilers: an interaction with protein content of feed on performance and the endocrine system. *Poultry Science* 77, 54-61.
40. Klinger, C.A., Gehad, A.E., Hulet, R.M., Roush, W.B., Lillehoj, H.S. & Mashaly, M.M. (2000) Effects of photoperiod and melatonin on lymphocyte activities in male broiler chickens. *Poultry Science* 79, 18-25.
41. Bacon, W.L. & Nestor, K.E. (1977) The effect of various lighting treatments or the presence of toms on reproductive performance of hen turkeys. *Poultry Science* 56, 415-420.
42. Bacon, W.L. & Nestor, K.E. (1980) Energy savings in turkey laying hens using an intermittent lighting program. *Poultry Science* 59, 1953-1955.
43. Bacon, W.L. & Nestor, K.E. (1981) Modification of an intermittent lighting program for laying turkey hens. *Poultry Science* 60, 482-484.
44. Siopes, T.D. & Pyrzak, R. (1990) Effect of intermittent lighting on the reproductive performance of first-year and recycled turkey hens. *Poultry Science* 69, 142-149.
45. Siopes, T.D. (1994) Critical day lengths for egg production and photorefractoriness in the domestic turkey. *Poultry Science* 73, 1906-1913.
46. Siopes, T.D. (1999) Intermittent lighting increases egg weight and facilitates early photostimulation of turkey breeder hens. *Poultry Science* 78, 1040-1043.
47. Siopes, T.D. (1983) Effect of intermittent lighting on energy savings and semen characteristics of breeder tom turkeys. *Poultry Science* 62, 2265-2270.
48. Dawson, A. (2001) The effects of a single long photoperiod on induction and dissipation of reproductive photorefractoriness in European starlings. *General and Comparative Endocrinology* 121, 316-324.
49. Bacon, W.L., Long, D.W., Kurima, K. & Chapman, D.P. (1994) Coordinate patterns of secretion of luteinising hormone and testosterone in mature male turkeys under continuous and intermittent photoschedules. *Poultry Science* 73, 864-870.
50. Siopes, T.D., Parkhurst, C.R. & Baughman, G.R. (1986) Intermittent light and growth performance of male turkeys from 2 to 22 weeks. *Poultry Science* 65, 2221-2225.
51. Noll, S.L., El Halawani, M.E., Waibel, P.E., Redig, P. & Janni, K. (1991) Effect of diet and population density on male turkeys under various environmental conditions. 1. turkey growth and health performance. *Poultry Science* 70, 923-934.
52. Gill, D.J. & Leighton, A.T. (1984) Effects of light environment and population density on growth performance of male turkeys. *Poultry Science* 63, 1314-1321.
53. Lewis, P.D., Perry, G.C. & Sherwin, C.M. (1998) Effect of intermittent light regimens on the performance of intact male turkeys. *Animal Science* 67, 627-636.
54. Hester, P.Y. & Kohl, H. (1989) Effect of intermittent lighting and time of hatch on Large Broad-breasted White turkeys. *Poultry Science* 68, 528-538.
55. Woodard, A.E., Wilson, W.O. & Abplanalp, H. (1962) Rhythm of lay in chickens as influenced by a 16 hour "day". *Poultry Science* 41, 1758-1762.
56. Bhatti, B.M. & Morris, T.R. (1978) Entrainment of oviposition in the fowl using light-dark cycles. *British Poultry Science* 19, 333-340.
57. Etches, R.J. (1977) The effects of ahemeral light and dark cycles on growth and sexual maturity in chickens. *Poultry Science* 56, 1039-1040.

58. Morris, T.R. (1978) The photoperiodic effect of ahemeral light-dark cycles which entrain circadian rhythms. *British Poultry Science* 19, 207-212.
59. Zimmerman, N.G. & Nam, C.H. (1989) Temporary ahemeral lighting for increased egg size in maturing pullets. *Poultry Science* 68, 1624-1630.
60. Foster, W.H. (1968) The effect of light-dark cycles of abnormal lengths upon egg production. *British Poultry Science* 9, 273-284.
61. Morris, T.R. (1973)The effects of ahemeral light and dark cycles on egg production in the fowl. *Poultry Science* 52, 423-445.
62. DEFRA (2002) Environment, in: *Code of recommendations for the welfare of livestock – laying hens.* p. 17. London, DEFRA Publications.
63. Byerly, T.C. & Moore, O.K. (1941) Clutch length in relation to period of illumination in the domestic fowl. *Poultry Science* 20, 387-390.
64. Lewis, P.D. & Perry, G.C. (1991) Oviposition time: correlations with age, egg weight and shell weight. *British Poultry Science* 32, 1135-1136.
65. Fox, S., Morris, T.R. & Jennings, R.C. (1971) The use of non-24-hour cycles to manipulate egg weight in pullets. *World's Poultry Science Journal* 27, 159.
66. Shanawany, M.M., Morris, T.R. & Pirchner, F. (1993) Influence of sequence length on the response to ahemeral lighting late in lay. *British Poultry Science* 34, 873-880.
67. Shanawany, M.M. (1992) Response of layers to ahemeral light cycles incorporating age at application and changes in effective photoperiod. *World's Poultry Science Journal* 48, 156-164.
68. Warren, D.C. & Conrad, R.M. (1939) Growth of the hen's ovum. *Journal of Agricultural Research* 58, 875-893.
69. Melek, O., Morris, T.R. & Jennings, R.C. (1973) The time factor in egg formation for hens exposed to ahemeral light-dark cycles. *British Poultry Science* 14, 493-498.
70. Shanawany, M.M., Sørensen, P. & Pirchner, F. (1993) Genotypic differences in speed and magnitude of response to ahemeral lighting. *British Poultry Science* 34, 881-886.
71. Nøddegaard, F. (1998) Oviposition patterns and plasma melatonin rhythms in response to manipulations of the light:dark cycle. *British Poultry Science* 39, 653-661.
72. Bhatti, B.M. & Morris, T.R. (1988) Model for the prediction of mean time of oviposition for hens kept in different light and dark cycles. *British Poultry Science* 29, 205-213.
73. Bhatti, B.M. & Morris, T.R. (1978) The effect of ahemeral light and dark cycles on patterns of food intake by the laying hen. *British Poultry Science* 19, 125-128.
74. Leeson, S. & Summers, J.D. (1985) Early application of conventional or ahemeral photoperiods in an attempt to improve egg size. *Poultry Science* 64, 2020-2026.
75. Cooper, J.B. & Barnett, B.D. (1976) Ahemeral photoperiods for chicken hens. *Poultry Science* 55, 1183-1187.
76. Leeson, S. & Summers, J.D. (1988) Significance of growing photoperiods and light stimulation at various ages for leghorn pullets subjected to regular or ahemeral photoperiods. *Poultry Science* 67, 391-398.
77. Nordstrom, J.O. & Ousterhout, L.E. (1983) Ahemeral light cycles and protein levels for older laying hens. *Poultry Science* 62, 525-531.
78. Shanawany, M.M. (1981) Sexual maturity and subsequent performance in the female domestic fowl. *PhD thesis*, University of Reading, UK.
79. Zimmerman, N.G. & Nam, C.H. (1989) Temporary ahemeral lighting for increased egg size in maturing pullets. *Poultry Science* 68, 1624-1630.
80. Morris, T.R. & Bhatti, B.M. (1978) Entrainment of oviposition in the fowl using bright and dim light cycles. *British Poultry Science* 19, 341-348.
81. Fitzsimmons, R.C. & Newcombe, M. (1991) The effects of ahemeral light-dark cycles early in the laying cycle on egg production in White Leghorn hens. *Poultry Science* 70, 20-25.
82. Leeson, S., Summers, J.D. & Etches, R.J. (1979) Effect of a 28-h light-dark cycle on egg shell quality of end-of-lay birds. *Poultry Science* 58, 285-287.

83. Nordstrom, J.O. (1982) Shell quality of eggs from hens exposed to 26- and 27-h light-dark cycles from 56 to 76 weeks. *Poultry Science* 61, 804-812.
84. Shanawany, M.M. & Pirchner, F. (1992) Performance of breeding flocks under ahemeral lighting. *Archiv für Geflügelkunde* 56, 227-229.
85. Spies, A.A., Robinson, F.E., Renema, R.A., Feddes, J.J., Zuidhif, M.J. & Fitzsimmons R.C (2000) The effects of body weight and long ahemeral days on early production parameters and morphological characteristics of broiler breeder hens. *Poultry Science* 79, 1094-1100.
86. Boersma, S.I., Robinson, F.E. & Renema, R.A. (2002) The effect of twenty-eight-hour ahemeral day lengths on carcass and reproductive characteristics of broiler breeder hens late in lay. *Poultry Science* 81, 760-766.
87. Siopes, T.D. & Neely, E.R. (1997) Ahemeral lighting of turkey breeder hens. 1. Cycle length effects on egg production and egg characteristics. *Poultry Science* 76, 761-766.
88. Siopes, T.D. & Neely, E.R. (1999) Ahemeral lighting of turkey breeder hens. 3. Temporary application and early age at lighting. *Poultry Science* 78, 387-391.
89. Siopes, T.D. & Neely, E.R. (1997) Ahemeral lighting of turkey breeder hens. 2. Early age at lighting and reproductive performance. *Poultry Science* 76, 1783-1788.
90. Kadono, H.& Usami, E. (1983) Ultradian rhythm of chicken body temperature under continuous light. *Japanese Journal of Veterinary Science* 45, 401-405.
91. Cain, J.R. & Wilson, W.O. (1974) The influence of specific environmental parameters on the circadian rhythms of chickens. *Poultry Science* 53, 1438-1447.
92. Beane, W.L., Siegel, P.B. & Siegel, H.S. (1965) Light environment as a factor in growth and feed efficiency of meat-type chickens. *Poultry Science* 44, 1009-1012.
93. Shutze, J.V., Jensen, L.S., Carver, J.S. & Matson, W.E. (1960) Influence of various lighting regimes on the performance of broiler chickens. *Washington Agricultural Experimental Station Technical Bulletin* 36, pp. 1-11.
94. Siegel, H.S. & Wood, G.W. (1964) Length of daily feeding time as an influence on growth and digestive efficiency in young chickens. *Poultry Science* 43, 406-410.
95. Siegel, P. B., Beane, W.L. & Kramer, C.Y. (1962) The measurement of feeding activity in chickens to 8 weeks of age. *Poultry Science* 41, 1419-1422.
96. Foshee, D.P., Centa, D.M., McDaniel, G.R. & Rollo, C.A. (1970) Diurnal activity patterns of broilers in a controlled environment. *Poultry Science* 49, 1514-1518.
97. Davis, G.S. & Siopes, T.D. (1985) The effect of light duration on turkey poult performance and adrenal function. *Poultry Science* 64, 995-1001.
98. Penquite, R. & Thompson, R.B. (1933) Influence of continuous light on Leghorns. *Poultry Science* 12, 201-205.
99. Lauber, J.K. (1964) Reproductive performance of domestic fowl maintained under continuous light. *Journal of Reproduction and Fertility* 7, 409-411.
100. Duncan, I.J.H. & Hughes, B.O. (1975) Feeding activity and egg formation in hens lit continuously. *British Poultry Science* 16, 145-155.
101. Savory, C.J. (1977) Effects of egg production on the pattern of food intake of broiler hens kept in continuous light. *British Poultry Science* 18, 331-337.
102. Morris, T.R. (1977) The clutch patterns of hens in constant illumination. *British Poultry Science* 18, 397-405.
103. Bhatti, B.M. (1987) Distribution of oviposition times of hens in continuous darkness or continuous illumination. *British Poultry Science* 28, 295-306.
104. Morris, J.A. (1961) The effect of continuous light and continuous noise on pullets held in a sealed chamber. *Poultry Science* 40, 995-1000.
105. Cherry, P. & Barwick, M.W. (1962) The effect of light on broiler growth. II. Light patterns. *British Poultry Science* 3, 41-50.
106. King, D.F. (1961) Effects of increasing, decreasing, and constant lighting treatments on growing pullets. *Poultry Science* 40, 479-484.

107. King, D.F. (1962) Egg production of chickens raised and kept in darkness. *Poultry Science* 41, 1499-1503.
108. Wilson, W.O. & Woodard, A.E. (1958) Egg production of chickens kept in darkness. *Poultry Science* 37, 1054-1057.
109. Morris, T.R., Fox, S. & Jennings, R.C. (1964) The response of laying pullets to abrupt changes in daylength. *British Poultry Science* 5, 133-147.

5 ILLUMINANCE (Light intensity)

Illuminance is synonymous with light intensity. This chapter describes its measurement, the appropriateness of the lux unit for defining illuminance for poultry, and the production and behavioural responses of the various types of poultry to illuminance.

• Measurement and units of illuminance •

Illuminance is the quantity of light falling on a unit area of surface and is the photometric equivalent of irradiance (W/m²). A light meter measures illuminance by converting the physical radiation output from a light source into a photometric measurement of the human perception of illuminance. The unit of measurement is the lux (page 5); though old light meters may still be calibrated in the imperial unit of foot-candles (page 3). The lux is calculated by measuring the radiant power (W) from a light source, dividing it by 12.57 (4π steradians in a sphere) to give the luminous intensity in candela, then multiplying by 683 to produce lumens, and by the spectral luminous efficiency for humans at each of the wavelengths of radiation (Table 1.2, Figure 5.1). The scaling factor is 683 because the luminous intensity (candela) at the maximum spectral luminous efficiency for humans (frequency of 540 x 10¹² Hertz ≡ a wavelength of 555nm) when the eye is adapted to bright light and retinal photo-reception is mainly by the cones (photopic conditions) is 1/683 W/sr (1). An example of a calculation of illuminance is given in Table 5.1.

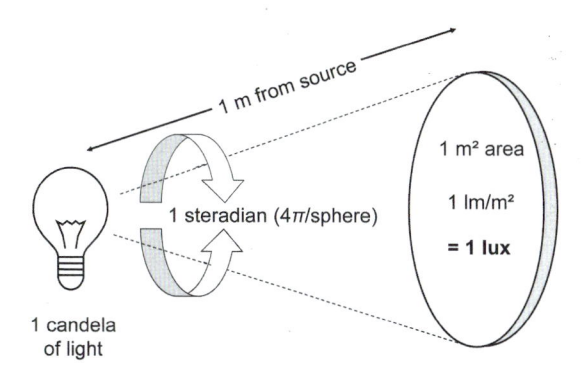

Figure 5.1 *The relationship between the radiant intensity from a light source in candela and illuminance in lux.*

Illuminance for poultry

The lux unit measures only the human perception of light intensity because it uses human, and not poultry, spectral luminous efficiencies in its computation. Thus the lux is not strictly an appropriate unit for measuring the fowl's perception of illuminance (2). Poultry spectral luminous efficiencies differ from those of humans (Table 1.2) because birds are sensitive to some ultraviolet (UV) radiation, which humans are not, and because poultry are more sensitive to the blue and red parts of the spectrum (refer to avian vision and comparative luminous efficiencies on page 7). A light meter takes no account of radiant output that is < 400 nm, and this underestimates the bird's response to blue and red light. However, the measurement of UV requires sophisticated equipment that cannot easily be used in poultry houses and the use of photocells calibrated for poultry would be prohibitively expensive; so light intensity in poultry houses is still measured in lux. Illuminance levels for domestic fowl can be found by entering the power output from the light source in 5-10 nm intervals (data available from lamp manufactures) and the fowl's spectral luminous efficiencies (Table 1.2) into a spreadsheet, and using the equation:

$$I = (w.s.683)/(12.566.d^2)$$

where w is the power output of the lamp (W) in 5-10 nm segments, s is the relative sensitivity of domestic fowl, 683 is the maximum luminous efficacy for human photopic vision (lumens/W), 12.566 (4π) is the number of steradians in a sphere, and d the distance from the lamp (m).

This unit has been variously called *clux* or *gallilux*. However, even this will only give an estimate of the bird's perception because the

Table 5.1 *Irradiance, luminous efficiency and illuminance at a distance of 1.8 m for a typical 7W compact warm-white fluorescent lamp, assuming a reflectance of 20% and using the spectral luminous efficiencies from Table 1.2.*

Wavelength (nm)	Power output (mW)	Lumens [1]	Galli-lumens [1]	Wavelength (nm)	Power output (mW)	Lumens [1]	Galli-lumens [1]
310	0.3	0.0	0.0	555	12.9	8.8	8.7
320	2.6	0.0	0.0	560	6.6	4.5	4.5
330	1.5	0.0	0.0	565	6.1	4.1	4.1
340	3.2	0.0	0.0	570	5.7	3.7	3.8
350	3.0	0.0	0.0	580	36.8	21.9	22.3
360	3.0	0.0	0.1	590	60.7	31.4	27.4
370	17.0	0.0	1.7	600	47.9	20.6	17.3
380	2.7	0.0	0.4	610	187.0	64.3	70.1
390	1.8	0.0	0.2	620	105.2	27.4	40.7
400	1.2	0.0	0.1	630	64.9	11.7	29.2
410	32.8	0.0	3.5	640	13.4	1.6	5.0
420	4.1	0.0	0.6	650	15.2	1.1	3.4
430	6.2	0.1	1.3	660	11.9	0.5	1.7
440	78.4	1.2	23.8	670	8.8	0.2	1.0
450	9.8	0.3	3.7	680	5.6	0.1	0.5
460	9.3	0.4	4.2	690	8.7	0.0	0.6
470	8.5	0.5	4.3	700	4.8	0.0	0.2
480	8.5	0.8	4.8	710	26.2	0.0	0.8
490	43.7	6.2	24.6	720	6.6	0.0	0.1
500	19.3	4.3	9.0	730	1.2	0.0	0.0
510	4.6	1.6	2.0	740	0.9	0.0	0.0
520	2.7	1.3	1.4	750	1.0	0.0	0.0
530	3.3	1.9	1.9	760	1.9	0.0	0.0
540	79.6	51.9	50.8	770	0.6	0.0	0.0
550	187.9	127.7	124.8	780	0.2	0.0	0.0
				Total	1176	400	505

Irradiance = 1176/(4π*1.8m*1.8m) = 28.9 W/m² (a sphere contains 4π steradians, the spherical area for a steradian equals the square of the radius or distance from the lamp, see Figure 5.1)
Lamp efficiency = 400 lumens/7W = 57.1 lm/W
Illuminance:
Human = 400*1.20 reflectance/(4π*1.8m*1.8m) = 11.8 lux
Poultry = 505*1.20 reflectance/(4π*1.8m*1.8m) = 14.9 *gallilux* or *clux*

[1] = power output (mW) x 0.683 x luminous spectral efficiency (1 W=683 lumens, 1 mW=0.683 lumens)
e.g. 490nm = 43.7 x 0.683 x 0.208 = 6.2 lumens, and 43.7 x 0.683 x 0.824 = 24.6 *gallilumens*

683 lumens/W scaling factor is a figure determined for humans. If the perception of brightness for a given intensity of natural light were the same for poultry and humans, the conversion factor would be about 400 and light meters would be over-stating the bird's perception of illuminance in lighting environments that have no UV radiation (further discussion on page 8).

Illuminance and wavelength
When voltage reduction equipment is used to reduce the intensity of white light, it also alters its spectral composition by increasing the proportion of longer wavelengths of light thus making the light redder. For example, the use of voltage reduction equipment to reduce illuminance from 25 to 3 lux increases the proportion of red light (630-780 nm) from 0.69 to 0.78 of visible light (3). This confounding of illuminance with spectral output may explain some of the anomalies that have been noted from time to time in reports of light intensity trials. The correct procedure for dimming is to use different wattage lamps, physical shielding or neutral filters, although these filters should not be used when the light source is emitting UV radiation because they are commonly made of polyester which will itself filter out UV.

• Growing pullets: effects on sexual maturation •

The effect of illuminance on sexual development was first investigated in early genotypes of domestic fowl using 0.2, 1.0 or 5.0 lux throughout a 6-h photoperiod (4). This showed that there was little effect on sexual maturity between 1.0 and 5.0 lux, but a significant delay compared with the other two intensities when pullets were reared under 0.2 lux light (○ in Figure 5.2). Subsequently, the effects were studied in modern brown-egg hybrids using six intensities between 0.05 and 2.2 lux, but with a 3-h period of dim light given immediately before and after an 8-h photoperiod of normal white incandescent light of between 5.5 and 10.8 lux (5). The purpose of this design was to find a threshold illuminance above which the pullet would treat supplementary dim light as an extension of the 8-h photoperiod and form a 14-h day, and below which it would ignore the dim light and respond only to the 8 h of normal illumination. It was concluded that the threshold illuminance at the feed trough (in cages) for stimulation of the photoperiodic mechanism lies between 0.9 and 1.7 lux (Figure 5.2) and that, below this threshold, very dim light (0.9 > 0.05 lux) may shift the photoinducible phase (page 13) so that a normal 8-h photoperiod becomes mildly stimulatory and advances age at first egg (AFE) by about 10 d.

Results from a trial that studied the effects of a change in light intensity during the rearing period in pullets maintained on a 10-h photoperiod showed that there are small differences between genotypes in their response to changes in intensity. Mean AFE, which was similar for constant 3 and 25-lux controls in brown-egg hybrids, was 3 d later for white-egg hybrids held at 3 lux than at 25 lux (3). However, such a delay will not be a major concern for the commercial poultry industry, especially in comparison to the increased electricity costs incurred in providing the brighter intensity. Surprisingly, sexual maturity was delayed by an increase in illuminance from 3 to 25 lux at both 9 and 16 weeks, but

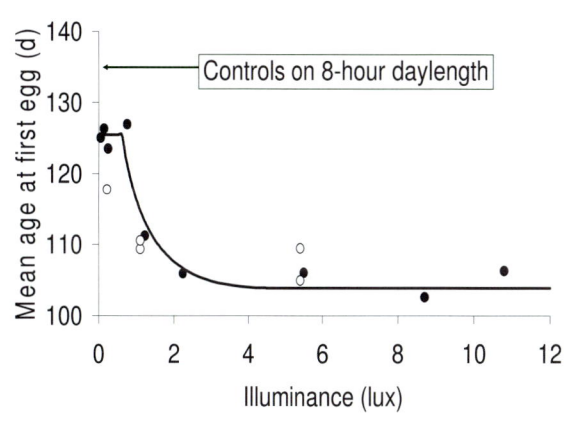

Figure 5.2 *Mean age at first egg in pullets reared on 6-h photoperiods at 0.2, 1.0 or 5.0 lux (○), or given 3 h of supplementary dim light immediately before and after a normally illuminated 8-h photoperiod at various intensities between 0.05 and 10.8 lux (●), or maintained on normally illuminated 8-h photoperiods (4, with data adjusted by least squares, and 5).*

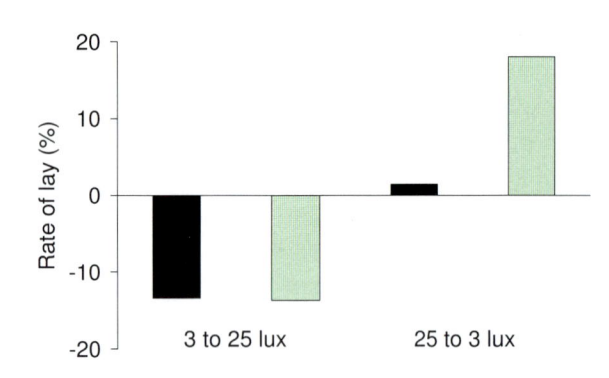

Figure 5.3 *Mean rate of egg production at 140 d for brown- and white-egg pullets, relative to that for pullets maintained at the final illuminance, for transfers from 3 to 25 lux or 25 to 3 lux at 9 weeks (black bars) or 16 weeks (grey bars) of age (3).*

Table 5.2 *Effect of an illuminance of 3 or 25-lux on feed intake and body weight in brown- and white-egg hybrids maintained on 10-h photoperiods (3).*

	White-egg hybrids		Brown-egg hybrids	
Intensity (lux)	3	25	3	25
Feed intake (kg)				
0 to 9 weeks	2.19	2.11	2.34	2.32
9 to 16 weeks	3.22	3.17	3.71	3.63
16 to 20 weeks	1.73	1.84	2.08	2.19
0-20 weeks	7.14	7.12	8.13	8.14
Body weight (kg)				
9 weeks	0.79	0.74	0.91	0.89
16 weeks	1.25	1.22	1.56	1.53
20 weeks	1.49	1.46	1.86	1.84

advanced by a decrease from 25 to 3 lux, though only for a change at 16 weeks (Figure 5.3, 3).

It was suggested than either the opposing changes in illuminance differentially modified the amplitude of the phase response curve for photoinducibility, and thereby shifted the photoinducible phase, or that differences in the ratio of retinally to intra-cranially received light resulted in a more robust photosexual response when the pullets were transferred to dimmer light and the reverse when changed to brighter light (3). Evidence to support the second hypothesis is the earlier maturity of genetically blind (6) and enucleated (7) birds, and the suppression of luteinizing hormone release by retinally received light (8).

Pullets maintained on 10-h photoperiods at 3 lux had higher feed intakes to 9 and 16 weeks of age than pullets exposed to 25 lux, but smaller appetites between 16 and 20 weeks, resulting in similar cumulative intakes to 20 weeks. As a consequence, pullets given 3-lux light were significantly heavier than those given 25 lux at 9 weeks, with the difference in body weight persisting through to 16 and 20 weeks (Table 5.2, 3). The suppressive effects of brighter illuminance to 9 weeks on feed intake and growth agree with the effects of illuminance on growth in broilers (9), feed intake in laying hens (10) and on both traits in turkeys (11).

• Laying hens: effects on laying performance •

Rate of lay

A modern review of the effects of illuminance on rate of lay in egg-type hybrids (10) shows that there is no need to change the optimum light intensity from the previously recommended 5 lux (4, 12). Interestingly, these findings agreed with conclusions made more than 60 years ago for the responses of laying hens to illuminance (13, 14). However, an analysis (10) of the three most recent sets of data (15, 16, 17) revealed that the response of modern genotypes to illuminance is described by the equation:

$$y = 0.817 + 0.0085I$$

where y = mean rate of lay (eggs/bird.d) and I = mean illuminance at the feed trough ($\log_{10}$ lux).

A comparison of predicted changes in rates of lay using this equation and that of an earlier analysis (12) suggests that the consequences of exposing modern egg-type hybrids to an illuminance below the recommended optimum of 5 lux are less damaging than for earlier stock. For example, between 1 and 5 lux, the earlier model predicted a reduction of 2.5 eggs/100 bird.d in rate of lay, whilst egg production for modern hybrids would only be expected to fall by 0.6 eggs/100 bird.d (Figure 5.4). It is clear

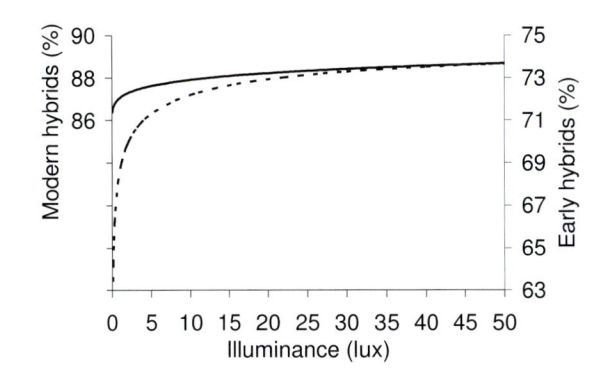

Figure 5.4 *A comparison of the regressions of mean rate of lay (%) on illuminance at the feed trough (lux) for early (broken line) and modern (solid line) egg-type hybrids (10).*

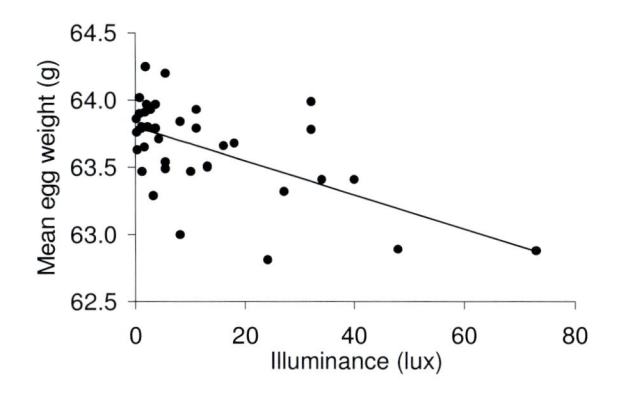

Figure 5.5 *The regression of mean egg weight on illuminance at the feed trough (lux) for egg-type hybrids (10).*

that modern genotypes are able to produce close to their potential when light intensity at the feed trough is as low as 0.5 lux and this, together with the evidence of excellent performance when hens have been provided with only 8-h daylengths (18), strongly suggests that the need for photostimulation in laying hens has diminished.

Illuminance during the rearing period seems to have no effect on subsequent rate of lay provided illuminance in the laying period is optimal (4).

Egg weight
Up to about 70 lux, mean egg weight decreases linearly by 0.13 g for each 10-lux decrease in illuminance (Figure 5.5, 10). This is described by the equation:

$$y = 63.8 - 0.0127I + k$$

where y = mean egg weight (g), I = mean illuminance at the feed trough (lux) and k = a constant for differences between genotypes.

It is not clear why egg weight decreases with increasing illuminance, but the effect is so small that investigation of the underlying mechanism would be very difficult.

Feed intake
The relationship between feed intake and illuminance is shown in Figure 5.6 (10). This is described by the equation:

$$y = 122.5 - 0.018I + k$$

where y = mean daily feed intake (g), I = mean illuminance at the feed trough (lux) and k = a constant for differences between genotypes.

In none of the six data sets used to construct this figure were the differences between treatments significant. However, a meta-analysis, with mean differences between data sets removed by least squares, shows that there is a significant negative association of feed intake and illuminance (10).

There are positive relationships of illuminance with egg mass output (10), energy expenditure and activity (19), and so the smaller appetites at brighter intensities are unexpected and difficult to explain, especially when laying hens spend less time eating and consume feed at a slower rate with an illuminance of < 1 lux than 6, 20 or 200 lux (20).

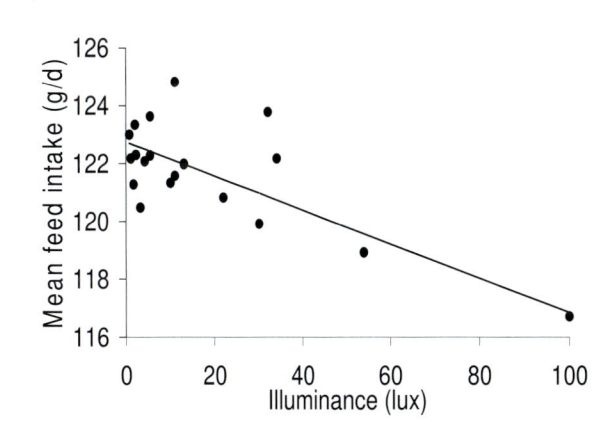

Figure 5.6 *A regression of mean feed intake during the laying period on illuminance at the feed trough for egg-type hybrids (10).*

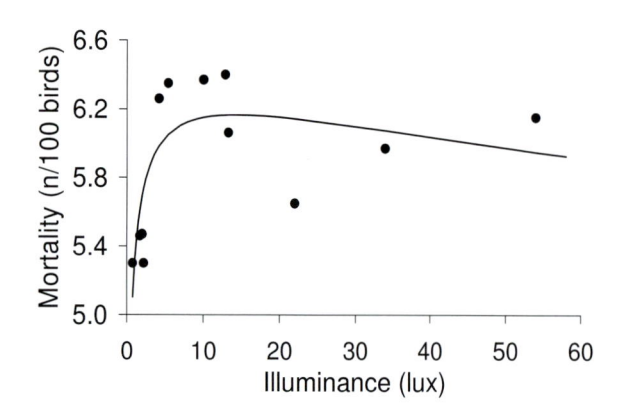

Figure 5.7 *A regression of mortality (n/100 birds) during the laying period on illuminance at the feed trough (lux) for egg-type hybrids housed in cages (10).*

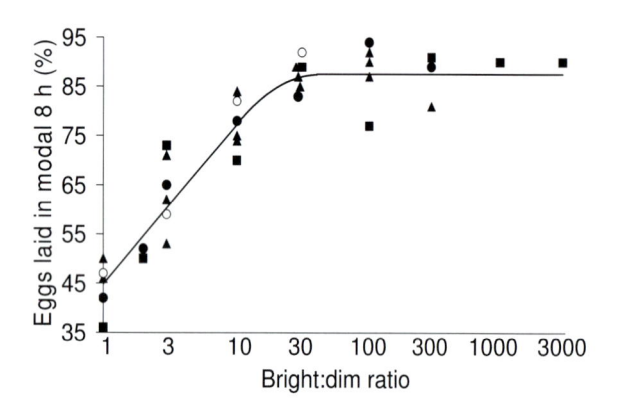

Figure 5.8 *Percentage of eggs laid in the modal 8 h by hens exposed to various ratios of bright:dim light within a 25 or 27-h cycle using a bright period of 10-lux (○), 30-lux (●), 100-lux (■) or 270-lux (▲) (24).*

Mortality

Incidences of mortality are not always given in reports of experiments that have studied the laying hen's response to illuminance. Analysis of four data sets (15, 16, 17, 21) suggests that mortality increases between 2 and 5 lux, but that higher intensities have no adverse effect on liveability (Figure 5.7).

Although none of the reports gave causes of death, the 2-lux threshold, above which mortality rates increase, agrees with a recommendation that illuminance should not exceed 2 lux for the satisfactory control of cannibalism in laying hens (22).

Oviposition entrainment

Egg-laying by hens given continuous illumination at 0.3 lux was spread throughout the 24 h, with a diurnal pattern similar to that of hens given continuous darkness (see Figure 4.11, page 73, 23). However, the provision of 1 h of light at 5 lux and 23 h at 0.3 lux was sufficient to induce complete entrainment, with more than 80% of the eggs laid in the modal 8 h, although the mean oviposition time (MOT) was more than 3 h earlier than for hens given 1 h of light at 5 lux and 23 h of darkness.

When hens were continuously lit with two light intensities alternating for various cycles of between 21 and 30 h, the degree of entrainment of egg-laying was dependent upon the ratio between the intensities but independent of the absolute illuminance (Fig 5.8, 24). It is clear that in the absence of a dark period, hens treated the brighter of the two photoperiods as day and the dimmer period as night. However, the bright:dim ratio required to achieve full entrainment varies with the cycle length from 10:1 for cycles of between 24 and 27 h to 1000:1 for 30-h cycles (see Figure 4.9, page 69 (24)).

When laying hens are given a mixture of bright light, dim light and darkness, the effect of the dim light depends on whether it is placed before or after the bright photoperiod, and on the absolute illuminance. When 8 h of dim light (0.06-0.12 lux) was given before an 8-h period of bright light (4-10 lux), the pattern of egg-laying was similar to that for hens given a 16L:8D schedule, though MOT was delayed by 2-3 h. However, when the 8 h of dim light was given after the main photoperiod, the hourly distribution of egg-laying and MOT were not significantly different from that of hens given an 8L:16D schedule (25). In contrast, when the dim period had an illuminance of 1.25 or 5 lux and followed a 20-lux main photoperiod, the pattern of egg-laying and MOT were similar to those for 16L:8D controls (23). The threshold for stimulation of the photoperiodic mechanism, at least for that controlling sexual maturation, lies somewhere between 0.9 and 1.7 lux (Figure 5.2, 5); in the first trial the dim period was below and, in the second trial, above this range.

However, the rules for entrainment are not the same as the rules for photoperiodic stimulation. Very dim light was perceived as dawn when it preceded the main photoperiod but was treated as darkness when it followed the normal day.

Activity and energy expenditure

Activity and activity-related energy expenditure are positively correlated with light intensity, and increase with the logarithm of illuminance (19). The simple relationship between illuminance and energy expended in activity is described by the equation:

$$y = 2.609 + 1.757I$$

where y = heat production (J/h.kg$^{0.75}$), and I = illuminance (log$_{10}$ lux). However, the data indicate a plateau between about 5 and 15 lux, with heat output rising at about 2.7 kJ/h.kg$^{0.75}$ per log$_{10}$ lux of illuminance above and below the plateau (Figure 5.9, 19). This plateau coincides

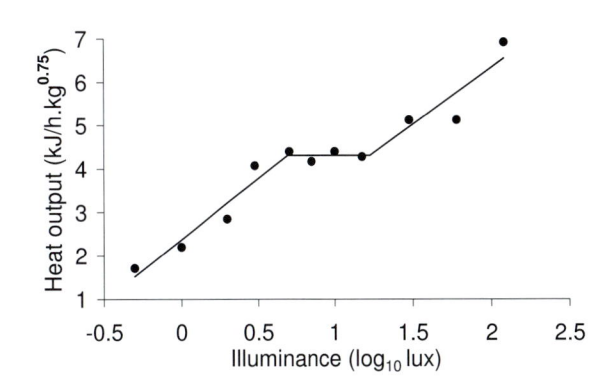

Figure 5.9 *Activity-related energy expenditure (kJ/h.kg$^{0.75}$) in laying hens exposed to various light intensities between 0.5 and 120 lux (19).*

with the range in which rate of lay is optimised (Figure 5.4), and so 10 lux seems to be the ideal illuminance in terms of egg production and feed conversion efficiency, as well as complying with welfare recommendations. Resting energy expenditure is independent of illuminance.

• Broiler breeders •

Although there is a dearth of experimental data on the responses of broiler breeders to illuminance, light intensity recommendations from the major broiler breeding companies are in reasonable agreement. Presumably these are supported by a wealth of practical experience. Most companies advise about 60 lux for the first 2 or 3 d followed by a progressive reduction to reach 5-10 lux by 1 to 3 weeks. This intensity is then maintained until 20 weeks. Recommendations for the laying period range from 15-20 lux up to ≥ 40 lux.

Recent work shows that the response of broiler breeders to photoperiod is similar to that of turkeys, presumably because, unlike egg-type hybrids, they both exhibit photorefractoriness. There have been several studies of the effect of illuminance on rate of lay in turkeys, and a meta-analysis of 12 sets of data (Figure 5.14, page 88), with differences between sets removed by least squares, suggests that, if the two species respond similarly, breeding company recommendations are correct for optimising egg production in broiler breeders. A comparison of the response of early and modern egg-type pullets to illuminance showed that the early hybrids, which would be more like broiler breeders in terms of

Table 5.3 *Estimates of the effect of illuminance on egg production to 60 weeks in broiler breeders, assuming that they respond to illuminance similarly to turkeys.*

Illuminance (lux)	Eggs to 60 weeks	Change from 40 lux
5	136.2	-13.9
10	140.9	-9.2
20	145.5	-4.6
30	148.1	-2.0
40	150.1	*
50	151.6	+1.5
60	152.0	+1.9

reproductive potential, were more affected by suboptimal illuminance than modern hybrids (Figure 5.4), and so broiler breeders may require a brighter illuminance in the laying period than the 5-10 lux recommended for egg-type birds. The cost of providing a brighter light intensity increases proportionately and these higher costs of electricity must be set against the value of the estimated changes in egg production (Table 5.3). A high proportion of broiler breeders are kept on litter, and so it will also be necessary to set illuminance at a level that dissuades hens from laying their eggs on the floor.

• Broilers •

In an early review of the effects of light intensity on broiler growth, it was concluded that body weight at 8-10 weeks was depressed by about 10 g for each doubling of the intensity (9). In each of six trials reported since 1988, it was concluded that illuminance had no significant effect on growth to between 6 and 9 weeks (26, 27, 28, and personal communication: S. H. Gordon). However, a meta-analysis of the modern data, which tested intensities between 0.1 and 200 lux, indicates that there is still a small, but significant, linear depression of growth, although the reduction between 1 and 100 lux is only about 20 g compared with 65 g for early genotypes (Figure 5.10). There is also a tendency for feed intake to decrease linearly with increasing illuminance, but by only 30 g between 1 and 100 lux (Figure 5.10).

There is also a tendency ($P=0.06$) for feed conversion efficiency to deteriorate as illuminance increases, but this is correlated with the logarithm of illuminance and, between 1 and 100 lux, only results in the consumption of 50 g more feed to produce a 2-kg bird.

Overall mortality, deaths due Sudden Death Syndrome and culling of birds with leg problems have been reported to be unaffected by light intensity in most investigations, despite the encouragement of activity by brighter illuminance (27). The relative effect of light intensity on performance traits appears to be similar for both sexes.

Cycling illuminance

The provision of a 12-h photoperiod at 40 lux, an 11-h period at 4 lux and a 1-h scotoperiod (29), and alternating between 5 and 100 lux at 2-hourly intervals without darkness (30) alter the activity patterns in broilers, but neither regime appears to benefit bird welfare or reduce the incidence of leg abnormalities.

Alternating 12-h bright and 12-h dim photoperiods also modifies the diurnal pattern of feeding activity of 4-week old broilers, depending on whether the feed is provided during the dim or during the bright phase (31). When fed during the bright period (94 lux), the birds learned to anticipate the ensuing period without feed and increased feeding activity during the final 2 h of the period, as they do in a normal light/dark schedule, but when the feed was provided during the dim phase, the birds were less able to anticipate the period without feed, but feeding during the first 2 h of the dim phase was enhanced (Figure 5.11).

Bird preference

When chickens are given a 20-h photoperiod and a choice of illuminance, the time and activity spent in each varies with age. In an investigation of the preference of broilers for a 6, 20, 60 or 200 lux environment, the birds spent most time in the brightest illuminance and least time in the dimmest at 2 weeks of age, but most time in the dimmest with equal periods in the other three environments at 6 weeks (Figure 5.12, 32). The birds spent between 60 and 70% of their time resting or perching, and it was the change with age in the choice of environment for periods of inactivity that accounted for the switch from generally preferring the brightest at 2 weeks to the dimmest at 6 weeks. Feeding, drinking, locomotion and litter-directed activities continued to be performed increasingly more often in brighter environments.

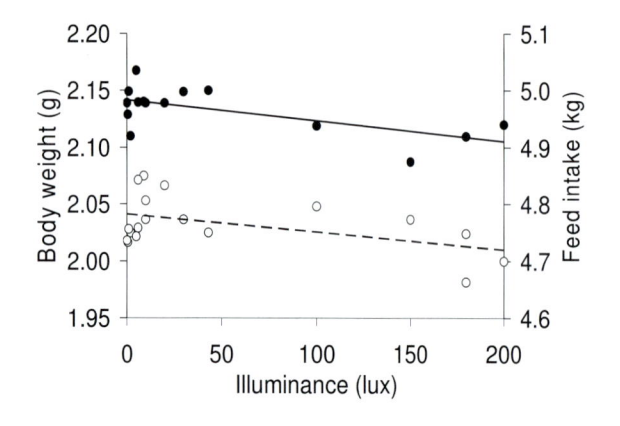

Figure 5.10 *Effect of illuminance on body weight at (●, solid line), and feed intake to (○, dotted line), 49 d. Meta-analysis of 6 sets of data (26, 27, 28 & personal communication: S. H. Gordon).*

Chapter 5 Illuminance

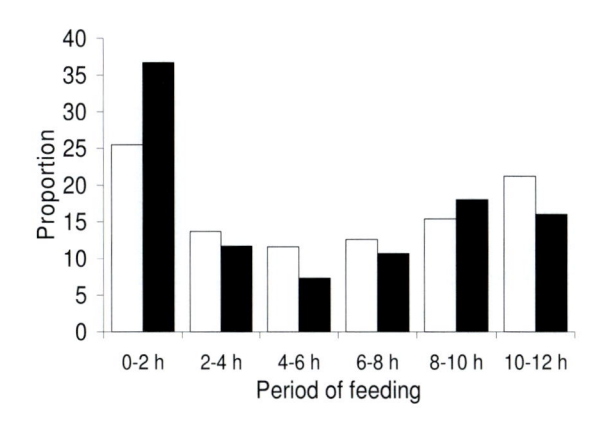

Figure 5.11 *Proportion (%) of daily feed intake consumed in 2-h periods of a 12-h feeding period by broilers given 12 h at 94 lux and 12 h at 16 lux (□ fed during bright phase, ■ fed during dim phase) (29).*

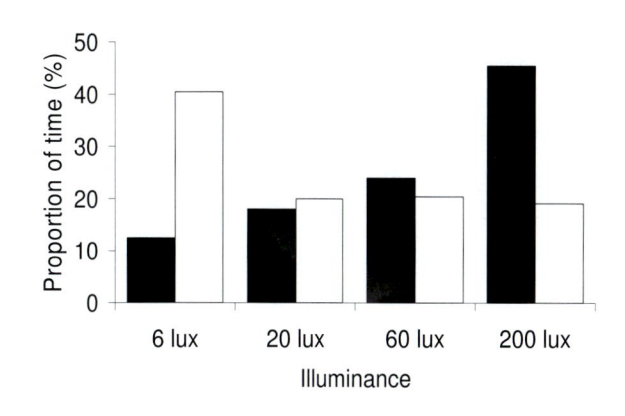

Figure 5.12 *Mean occupancy by chickens of a 6, 20, 60, or 200 lux environment at 2 (■) and 6 (□) weeks of age (32).*

• **Breeding turkeys** •

Pre-lay 'dark' period

Turkeys do not become fully responsive to a stimulatory photoperiod until they have dissipated juvenile photorefractoriness, and this requires at least 8 weeks of short days. However, it seems that complete photosensitivity cannot be gained unless the turkeys are housed in fully lightproof accommodation during this period. Many turkeys are kept in curtain-sided houses (brown-out conditions), and so, to investigate how light-tight buildings need to be, turkeys were given 6-h photoperiods at 650 lux, followed by 18 h at various intensities between 0.5 and 5 lux from 24 to 32 weeks. Whereas a control group given 18 h of true darkness reached 50% lay within 35 d of photo-stimulation, none of the 'brown-out' groups had reached 50% lay by the time the controls had been in production for 20 weeks. Each experimental group produced significantly fewer eggs than controls, with the depression of performance being inversely proportional to the 'dark-phase' illuminance (Table 5.4, 33). These findings, which show that the 'dark' phase needs to be below 0.5 lux to satisfactorily dissipate photorefractoriness, are supported by observations that when the groups were transferred back to a 6L:18Dim regimen, using the original levels of illuminance during the dim period, egg production was only terminated in

Table 5.4 *Effect of illuminance during the 'dark' (brown-out) phase of a 6L:18Dim pre-lay regimen on subsequent egg production in Large White turkeys and on rate of lay 8 weeks after the birds had been returned to 6L:18Dim lighting (33).*

Illuminance in 18-h dim period (lux)	Eggs/bird in 20 weeks' production	Rate of lay 8 weeks after return to 6L:18Dim (%)
0	50.4±8.1	0
0.5	22.3±4.7	12.0±4.2
1.0	22.3±1.8	43.4±5.4
1.5	15.7±3.4	38.0±4.7
2.0	17.3±4.0	44.6±6.6
3.0	17.3±2.7	45.1±5.2
5.0	11.4±3.6	43.4±4.5

the hens that were given 18 h of true darkness (Table 5.4, 33).

Recycling turkey hens with low illuminance

In some countries it is normal practice to keep turkey hens for more than one season, and photosensitivity is generally restored by returning the birds to short days for about 8 weeks. However, it is possible to use low light intensity to dissipate adult photorefractoriness,

Table 5.5 *Percentage of birds resuming egg production and mean rate of lay in 10 weeks for turkeys transferred back to 16 h at 55 lux after being given 8 weeks at 0.5, 2.2, 4.3 or 7.6 lux or a temporary transfer to 6-h photoperiods at 55 lux (35).*

8-week recycling lighting treatment		Proportion resuming egg-laying	Rate of lay (eggs/100 bird.d)
(h)	(lux)		
6	55	1.00	58.4
16	0.5	1.00	54.4
16	2.2	0.44	24.0
16	4.3	0.38	18.7
16	7.6	0.11	9.6

even when birds are maintained on stimulatory daylengths (34, 35). A reduction from 55 to 0.5 lux was as effective as an 8-week period of short days in restoring photosensitivity in turkeys kept on 16-h photoperiods (Table 5.5, 35). However, more than half of hens that had been transferred to 2.2, 4.3 or 7.6 lux failed to recommence egg-laying, indicating a threshold for dissipating adult photorefractoriness in long-day turkeys of between 0.5 and 2.2 lux. Nevertheless, the threshold varies from bird to bird, because the time to first egg in each of the five treatment groups and the mean rate of lay for the individuals that did resume egg production were similar in all groups.

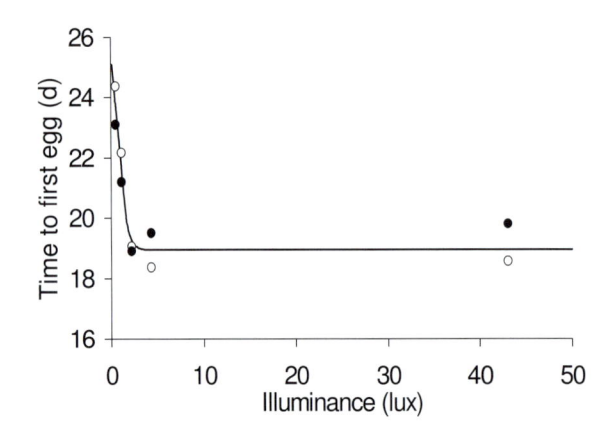

Figure 5.13 *Effect of illuminance on mean time of first egg following a transfer from 6 to 16-h photoperiods in Large White turkeys (35).*

Threshold for initiating sexual maturation

Although the intervals between photo-stimulation and mean age at first egg in Figure 5.13 indicate that that the threshold for optimising sexual maturation is about 2 lux, sexual maturity for turkeys transferred to long days at 0.5 lux will only be delayed by about 5 d (35). The 2-lux threshold for full inducement of sexual maturation agrees with the figure for egg-type pullets (see Figure 5.2, page 81) and is similar to the thresholds reported for English sparrows (36) and Bobwhite quail (37). It is possible, therefore, that this is a common threshold for initiating rapid gonadal development in all birds.

Illuminance for egg production

An analysis of egg production data from 12 trials shows that the optimal light intensity for maximising egg production in turkeys is between 50 and 60 lux (Figure 5.14, 33, 34, 38-43). This agrees with the conclusion from one of the trials that hens on 54 lux laid as many eggs as birds on 108, 216 or 324 lux (42). Up to 60 lux, mean rate of lay increases in proportion to the logarithm of the intensity, and is described by the equation:

$$y = 43.3 + 5.311I$$

where y = mean rate of lay (eggs/bird.d), I = mean illuminance at the feed trough ($\log_{10}$ lux). There is no perceptible increase in lay above 60 lux, but this is much brighter than both the 2 lux required to induce sexual maturation (Figure 5.13) and the 5-lux recommended for optimising

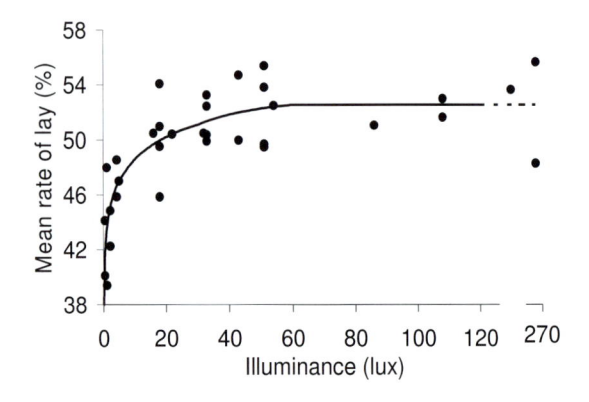

Figure 5.14 *Effect of illuminance on mean rate of lay (eggs/100 bird.d) in turkeys (33, 34, 38-43).*

egg production in domestic hens (see Figure 5.4, page 83). Modern domestic hens will produce about 1% more eggs at 60 lux than at 5 lux, and this increase will not normally be sufficient to offset the costs of providing the brighter illuminance. In contrast, turkey hens can be expected to increase their mean rate of lay from 47 to 53 eggs/100 bird.d (a 12% increase) when intensity is increased from 5 to 60 lux, and this larger gain and the extra value of a turkey hatching egg compared with an eating egg makes the provision of higher illuminance for turkeys economically viable.

Illuminance has generally had no effect on egg weight, feed intake, fertility or hatchability in turkeys. Increases in the number of poults produced have, without exception, been a consequence of better egg production and not improved hatchability (38, 39, 42, 43). Fertility in males, and subsequent hatch of fertile eggs, was reported to be similar at 5 and 43 lux (44). Likewise, illuminance had no significant effect on semen volume, sperm concentration or sperm numbers per ejaculate.

First and second-year turkeys respond similarly to light intensity (43).

• Growing turkeys •

Provided the light intensity during the first 1-2 weeks is at least 10 lux, illuminance generally has minimal effect on growth, feed intake or feed conversion efficiency in turkeys (45-49). However, in trials comparing 1, 11 and 110 with 220 lux, and 11 with 108 lux, there was a trend, though non-significant, for male body weight at 22 weeks to increase with illuminance and, in both trials, for body weight to be related to testicular weight (Figure 5.15, 47, 48). This is not surprising, because photoperiod can affect growth in male turkeys through a modification of testosterone release (see Figure 3.28, page 41).

When low intensity light (e.g. 1 lux) is given from 1 to 14 d, body weight gain is reduced, feed intake is depressed, mortality is markedly increased, and the eyes and adrenal gland are enlarged compared with brighter (≥ 10 lux) intensities (51).

If turkeys are beak-trimmed, de-snooded and toe-clipped, light intensity has little effect on mortality (47, 48). However, mortality and the need to cull injured birds increase with illuminance when male birds are left intact. The severity of the effect of light intensity on aggressive behaviour is increased when the birds are given longer photoperiods. For example, in a comparison of 8, 12, 16 and 23-h photoperiods at 1 or 10 lux from 1 d of age, it was necessary to reduce the illuminance from 10 to 1 lux at 75, 73 and 45 d respectively for the birds given 12, 16 and 23-h daylengths (49). In a separate study in which intact male turkeys were exposed to 5, 10, 36 or 70 lux, the incidence of injuries to the tail and wings up to 5 weeks was positively correlated with illuminance (51).

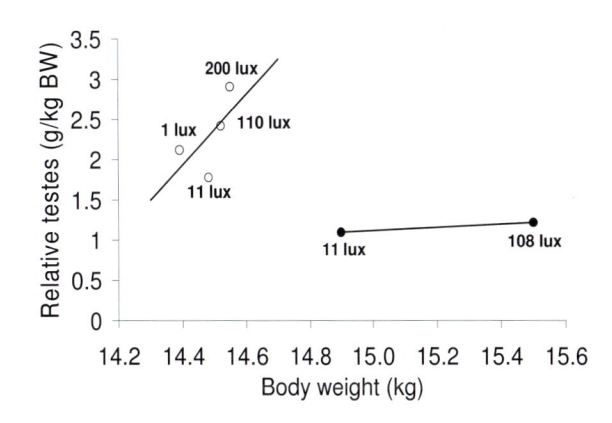

Figure 5.15 *Relationship between testicular weight (both testes) and body weight in turkeys exposed to 23-h photoperiods at 1, 11, 110 or 220 lux in one trial (○, 47) and to 11 or 108 lux in another trial (●, 48).*

• Waterfowl •

Geese

The effect of light intensity on the performance of geese has been variable. In a trial that studied the response to 18-h photoperiods made up of 10-12 h of natural light and artificial light at 20 or 50 lux, geese given the dimmer light were slower coming into lay but subsequently had better persistency and produced significantly more eggs (Figure 5.16, 52). In contrast, other experiments have reported similar performance when geese were given long days, in season, which comprised natural lighting and artificial light at 10, 20, 30 or 40 lux (53), or at 20, 120 or 220 lux when given out of season (54). The first trial was conducted in Israel and the other two in Taiwan. The timing of the breeding season in geese varies with latitude (55), and seasonal differences in daylength or illuminance between Israel and Taiwan might explain the contradictory findings.

Ducks

Egg-laying ducks

Egg-laying ducks are almost always given natural lighting, and so the response to light intensity will generally be irrelevant. However, where artificial light is used to supplement natural lighting, it is likely that, if ducks respond to combinations of bright light, dim light, and darkness in a similar way to domestic fowl (see page 84), the artificial light should precede the natural day because of the difference in illuminance. Unfortunately there are no data to support this suggestion.

Meat-type ducks

Transferring 22-week old Pekin breeder ducks from short days to a 16-h mixture of natural light and artificial light at 172 lux (high-pressure sodium lamp) or 10 lux (incandescent lamp) resulted in the brightly lit birds maturing 7 d earlier and having a superior rate of lay up to peak egg production (56). This suggests that ducks require more than 10 lux to maximise the initiation of sexual maturation, at least when it is given as a supplement in the evening after a natural dusk. Egg production was similar for both groups, but fertility was 2-3% higher for the 172-lux birds after peak lay. Egg weight and shell quality was unaffected by illuminance. Although the two light intensities were produced from different types of lamp, spectral differences were not thought to have contributed to the results.

In controlled environment conditions, however, where illuminance is constant throughout the photoperiod, 10 lux may be adequate for sexual development, as is the case for domestic fowl (see Figure 5.2, page 81) and turkeys (see Figure 5.13, page 88). If a duck's response to illuminance in the laying period matches early genotypes of laying hen (see Figure 5.4, page 83) and turkeys (see Figure 5.14, page 88), then illuminance in the breeding facilities might need to be about 50 lux to maximise egg numbers.

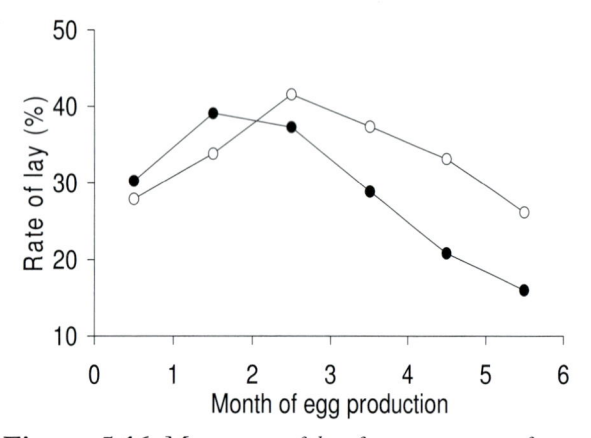

Figure 5.16 *Mean rate of lay for two groups of geese given 10 to 12 h of natural lighting augmented to an 18-h photoperiod with artificial lighting at an illuminance of 20 (○) or 50 (●) lux (52).*

• References •

1. Commission Internationale de l'Éclairage (1983) The basis of physical photometry. CIE, Vienna.
2. Nuboer, J.F.W., Coemans, M.A.J.M. & Vos, J.J. (1992) Artificial lighting in poultry houses: are photometric units appropriate for describing illumination intensities? *British Poultry Science* 33, 135-140.
3. Lewis, P.D., Sharp, P.J., Wilson, P.W. & Leeson, S. (2004) Changes in light intensity can influence age at sexual maturity in domestic pullets. *British Poultry Science* 45, 125-132.
4. Morris, T.R. (1967) The effect of light intensity on growing and laying pullets. *World's Poultry Science Journal* 23, 245-252.

5. Lewis, P.D., Morris, T.R. & Perry, G.C. (1999) Light intensity and age at first egg in pullets. *Poultry Science* 78, 1227-1231.
6. Ali, A. & Cheng, K.M. (1985) Early egg production in genetically blind (*rc/rc*) chickens in comparison with sighted (Rc^+/rc) controls. *Poultry Science* 64, 789-794.
7. Siopes, T.D. & Wilson, W.O. (1978) The effect of intensity and duration of light on photorefractoriness and subsequent egg production of chukar partridge. *Biology of Reproduction* 18, 155-159.
8. Yokoyama, K. & Farner, D.S. (1976) Photoperiodic responses in bilaterally enucleated female white crowned sparrows. *General & Comparative Endocrinology* 330, 528-533.
9. Morris, T.R. (1967) Light requirements of the fowl, in: *Environmental control in poultry production*. (Ed. T.C. Carter), pp. 15-39. Edinburgh, Oliver & Boyd.
10. Lewis, P.D. & Morris, T.R. (1999) Light intensity and performance in domestic pullets. *World's Poultry Science Journal* 55, 241-250.
11. Lewis, P.D., Perry, G.C. & Sherwin, C.M. (1998) Effect of photoperiod and light intensity on the performance of intact male turkeys. *Animal Science* 66, 759-767.
12. Morris, T.R. (1981) The influence of photoperiod on reproduction in farm animals, in: *Proceedings of 31st Nottingham Easter School in Agricultural Science*, pp. 458-461.
13. Roberts, J. & Carver, J.S. (1941) Electric lighting for egg production. *Agricultural Engineering* 22, 357-364.
14. Nicholas, J.E., Callenbach, E.W. & Murphy, R.R. (1944) Light intensity as a factor in the artificial illumination of pullets. *Pennsylvania Agricultural Experimental Station Bulletin* No. 462, pp. 1-24.
15. Hill, A.M., Charles, D.R., Spechter, H.H., Bailey, R.A. & Ballantyne, A.J. (1988) Effects of multiple environmental and nutritional factors in laying hens. *British Poultry Science* 29, 499-511.
16. Morris, T.R., Midgley, M.M. & Butler, E.A. (1988) Experiments with the Cornell intermittent lighting system for laying hens. *British Poultry Science* 29, 325-332.
17. Tucker, S.A & Charles, D.R. (1993) Light intensity, intermittent lighting and feeding regimen during rearing as affecting egg production and egg quality. *British Poultry Science* 34, 255-266.
18. Lewis, P.D., Perry, G.C., & Morris, T.R. (1997) Effect of size and timing of photoperiod increase on age at first egg and subsequent performance of two breeds of laying hen. *British Poultry Science* 38, 142-150.
19. Boshouwers, F.M.G. & Nicaise, E. (1987) Physical activity and energy expenditure of laying hens as affected by light intensity. *British Poultry Science* 28, 155-163.
20. Prescott, N.B. & Wathes, C.M. (2002) Preference and motivation of laying hens to eat under different illuminances and the effect of illuminance on eating behaviour. *British Poultry Science* 43, 190-195.
21. Skoglund, W.C., Palmer, D.H., Wabeck, C.J. & Verdaris, J.N. (1975) Light intensity required for maximum egg production in hens. *Poultry Science* 54, 1375-1378.
22. Wilson, W.O., Ernst, R.A., Thompson, J.F., Woodard, A.E. & Pfost, R.E. (1979) Lighting for Poultry. *Division of Agricultural Sciences, University of California* Leaflet 21067, pp. 1-14.
23. Bhatti, B.M., Mian, A.A. & Morris, T.R. (1988) Timing of oviposition in mixed systems using bright light, dim light and darkness. *British Poultry Science* 29, 395-401.
24. Morris, T.R. & Bhatti, B.M. (1978) Entrainment of oviposition in the fowl using bright and dim light cycles. *British Poultry Science* 29, 341-348.
25. Lewis, P.D., Perry, G.C., Morris, T.R. & English, J. (2001) Supplementary dim light differentially influences sexual maturity, oviposition time, and melatonin rhythms in pullets. *Poultry Science* 80, 1723-1728.
26. Newberry, R.C., Hunt, J.R. & Gardiner, E.E. (1986) Light intensity effects on performance, activity, leg disorders, and Sudden Death Syndrome of roaster chickens. *Poultry Science* 65, 2232-2238.
27. Newberry, R.C., Hunt, J.R. & Gardiner, E.E. (1988) Influence of light intensity on behaviour and performance of broiler chickens. *Poultry Science* 67, 1020-1025.

28. Charles, R.G., Robinson, F.E., Hardin, R.T., Yu, M.W., Feddes, J. & Classen, H.L. (1992) Growth, body composition, and plasma androgen concentration of male broiler chickens subjected to different regimens of photoperiod and light intensity. *Poultry Science* 71, 1595-1605.
29. Gordon, S.H & Tucker, S.A. (1996) Effect of light intensity on broiler welfare. *British Poultry Science* 37, S21-22.
30. Kristensen, H.H., Aerts, J. M., Leroy, T., Berckmans, D. & Wathes, C.M. (2004) Using light to control activity in broiler chickens. *British Poultry Science* 45, S30-31.
31. May, J.D. & Lott, B.D. (1994) Effects of light and temperature on anticipatory feeding by broilers. *Poultry Science* 73, 1398-1403.
32. Davis, N.J., Prescott, N.B., Savory, C.J. & Wathes, C.M. (1999) Preferences of growing fowls for different light intensities in relation to age, strain and behaviour. *Animal Welfare* 8, 193-203.
33. Siopes, T.D. (1991) Light intensity for turkey breeder hens: degree of darkness required during short-day light restriction and intensity threshold during photostimulation. *Poultry Science* 70, 1333-1338.
34. Marsden, S.J & Lucas, L.M. (1964) Effect of short day or low light intensity light treatments on reproduction of fall hatched turkeys in two environments. *Poultry Science* 43, 434-441.
35. Siopes, T.D. (1984) Recycling turkey hens with low light intensity. *Poultry Science* 63, 1449-1452.
36. Bartholomew, G.A. (1949) The effect of light intensity and daylength on reproduction in the English sparrow. *Bulletin of Mus and Comparative Zoology, Harvard University* 101, 433-476.
37. Kirkpatrick, C.M. (1955) Factors in photoperiodism of bobwhite quail. *Physiological Zoology* 28, 255-264.
38. McCartney, M.G. (1971) Reproduction of turkeys as affected by age at lighting and light intensity. *Poultry Science* 50, 661-662.
39. Nestor, K.E. & Brown, K.I. (1972) Light intensity and reproduction of turkey hens. *Poultry Science* 51, 117-121.
40. Thomason, D.M., Leighton, A.T. & Mason, J.P. (1972) A study of certain environmental factors on reproductive performance of Large White turkeys. *Poultry Science* 51, 1438-1449.
41. Siopes, T.D. (1984) The effect of high and low intensity cool-white fluorescent lighting on the reproductive performance of turkey breeder hens. *Poultry Science* 63, 920-926.
42. Siopes, T.D. (1991) Light intensity effects on reproductive performance of turkey breeder hens. *Poultry Science* 70, 2049-2054.
43. Siopes, T.D. (1992) Effect of light intensity level during prelay light restriction on subsequent reproductive performance of turkey breeder hens. *Poultry Science* 71, 939-944.
44. Jones, J.E., Hughes, B.L. & Wall, K.A. (1977) Effect of light intensity and source on tom reproduction. *Poultry Science* 56, 1417-1420.
45. Bacon, W.L. & Touchburn, S.P. (1976) Effect of light intensity on confinement rearing of male turkeys. *Poultry Science* 55, 999-1007.
46. Proudfoot, F.G., Hulan, H.W. & Dewitt, W.F. (1979) Response of turkey broilers to different stocking densities, lighting treatments, toe clipping and intermingling of the sexes. *Poultry Science* 58, 28-36.
47. Siopes, T.D., Timmons, M.B., Baughman, G.R. & Parkhurst, C.R. (1983) The effect of light intensity on the growth performance of male turkeys. *Poultry Science* 62, 2336-2342.
48. Siopes, T.D., Baughman, G.R., Parkhurst, C.R. & Timmons, M.B. (1989) Relationship between duration and intensity of environmental light on the growth performance of male turkeys. *Poultry Science* 68, 1428-1435.
49. Lewis, P.D., Perry, G.C. & Sherwin, C.M. (1998) Effect of photoperiod and light intensity on the performance of intact male turkeys. *Animal Science* 66, 759-767.
50. Siopes, T.D., Timmons, M.B., Baughman, G.R. & Parkhurst, C.R. (1984) The effects of light intensity on turkey poult performance, eye morphology, and adrenal weight. *Poultry Science* 63, 904-909.

51. Moinard, C., Lewis, P.D., Perry, G.C. & Sherwin, C.M. (2001) The effects of light intensity and light source on injuries due to pecking of male turkeys *(Meleagris gallopavo)*. *Animal Welfare* 10, 131-139.
52. Pyrzak, R.N., Snapir, N., Robinzon. B & Goodman, G. (1984) The effect of supplementation of daylight with artificial light from various sources and at two intensities on the egg production of two lines of geese. *Poultry Science* 63, 1846-1850.
53. Hsu, J.C., Peh, H.C. & Chen, Y.H. (1998) Effects of lighting program on the laying performance of geese. I. Effects of supplement of day light with artificial light of various light intensities on the laying performance of geese. *Journal of Agriculture and Forestry (Taiwan)* 39, 15-25.
54. Wang, C.M., Kao, J.Y., Lee, S.R. & Chen, L.R. (2005) Effects of artificial supplemental light on reproductive season of geese kept in open houses. *British Poultry Science* 46, 728-732.
55. Wang, S.D., Jan, D.F., Yeh, L.T., Wu, G.C. & Chen, L.R. (2002) Effect of exposure to long photoperiod during the rearing period on the age at first egg and the subsequent reproductive performance in geese. *Animal Reproduction Science* 73, 227-234.
56. Davis, G.S., Parkhurst, C.R. & Brake, J. (1993) Light intensity and sex ratio effects on egg production, egg quality characteristics, and fertility in breeder Pekin ducks. *Poultry Science* 72, 23-29.

Chapter 5 Illuminance

6 WAVELENGTH (Colour)

Humans have three types of cone in the retina of the eye, which have peak sensitivities at 450 nm, 550 nm and 700 nm. This allows us to perceive the primary colours of blue, green and red, and white when all cones are stimulated simultaneously. Avian eyes, in contrast, have a fourth type of cone with a peak sensitivity at about 415 nm and this, together with their optically clear lens and humours, allows poultry to be responsive to radiation below 400 nm and to 'see' in ultraviolet light.

This chapter describes the effects of wavelength, including ultraviolet radiation, on growth, reproduction, liveability and behaviour in poultry. In some areas there is a dearth of information relating specifically to domestic poultry, but we have assumed that most of the physiological processes involved in avian light reception and transmission are common to all species and that information from non-domesticated species can be safely applied to poultry.

• **Growth** •

Domestic fowl

When chicks have been exposed to various wavelengths of monochromatic light, but at the same irradiance ($W.m^{-2}$) or illuminance (lux), body weight gain to between 4 and 11 weeks has either shown a trend towards improvement (1-5) or has been significantly greater (6-8) in birds exposed to radiation of 415 to 560 nm (violet to green) than in birds given > 635 nm (red) or broad spectrum (white) light. However, it should be appreciated that, despite the similarity of irradiance or illuminance, the differences in retinal sensitivity across the visual spectrum could still have resulted in the birds perceiving the various wavelengths of light as different intensities. For example, although the irradiance of monochromatic LED and white incandescent lamps was equated to 0.1 $W.m^{-2}$ at head height in one trial (8), the birds would have perceived the intensity of 480, 560, 660 nm and white light as 56, 68, 15 and 26 gallilux, respectively. However, the heaviest body weights in this trial occurred at the wavelengths that the bird would have perceived as having the brightest intensity (480 and 560 nm), which makes it likely that the better growth was primarily a response to wavelength, not to intensity, because growth generally decreases with increasing illuminance (see page 86).

Turkeys

Similarly, body weight gain has been reported to be faster to 16 weeks in male and to 18 weeks in female turkeys when they have been exposed to blue light (450 nm) compared with red (650 nm) or white light at an illuminance of either 5 or 86 lux (9, 10). As the turkeys would have perceived the blue light as being brighter than the red or white light, the faster growth again occurred at the wavelength that was perceived as being the brightest. Since growth rate in turkeys is generally unaffected by illuminance between 5 and 86 lux (11), it seems likely that, as in the domestic fowl, the principal response was to wavelength and not intensity.

After 18 weeks, growth in male and female turkeys is faster under red and white than under blue light (9, 10). This is a period of rapid gonadal development in turkeys maintained on long daylengths and, since long wavelengths are more sexually stimulatory than short wavelengths (see page 11), the improved growth under red light is likely to be a response to increased plasma concentrations of sex steroids, rather than a direct effect of wavelength *per se* on growth. This is also the most likely explanation for the more efficient feed conversion recorded during this period by turkeys illuminated with red as opposed to blue light (9, 10).

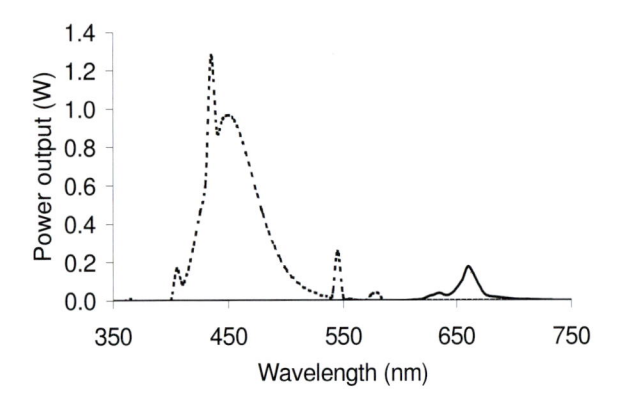

Figure 6.1 *Spectral power output (W/5nm) for typical blue (broken line) and red (solid line) fluorescent lamps.*

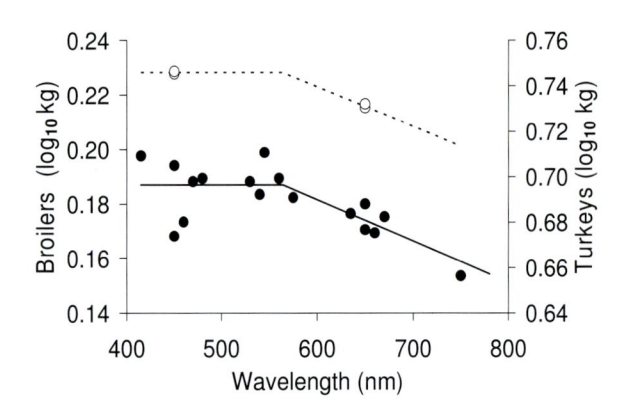

Figure 6.2 *Effect of wavelength of light on body weight (kg × log₁₀) in broilers (●) and turkeys (○) (2-10).*

Commercial lamps

Growth rates under different colours have not generally been significantly different where commercial lamps have been used (12-16); this does not necessarily disprove the hypothesis that long wavelengths suppress growth because most commercial coloured lamps are far from monochromatic (Figure 6.1). The anticipated better growth rate for birds maintained under a 'blue' or 'green' commercial light source, compared with white, might have been annulled by extraneous wavelengths of longer radiation.

General growth response to wavelength

A regression of 6 sets of broiler body weight data on wavelength between 415 and 750 nm showed a significant linear correlation, with the fastest growth consistently occurring in the blue-green region of the spectrum. However, it is unlikely that a linear regression is the appropriate model, because body weights under blue light have generally either been the same as, or inferior to, those under green light. Accordingly, a bent-stick model was applied to the data (transformed to log$_{10}$ values and adjusted for mean differences between trials) for comparisons where illumination had been equated to the same irradiance or illuminance and the mean body weight of the broilers was about 1.5 kg. This revealed a significant negative regression of body weight on wavelength between 530 and 750 nm (green to red light), decreasing by about 50 g for each 100 nm increase in wavelength, and an inflection at 564 nm (Figure 6.2). The model had an r² value of 0.997 and a residual standard deviation of 0.0071 log units. Only blue and red light have been compared in turkey trials, and so the appropriateness of a bent stick model could not be tested. However, because there were similar proportional changes in body weight for turkeys and broilers given blue compared with red light, a hypothetical bent stick model has also been fitted to the turkey data in Figure 6.2.

Whilst it might be pedantic to ask whether growth is better under blue and green light or poorer under red, the generally better growth under blue and green compared with white light, but similarity of growth under red to that under white light, suggests that growth is suppressed by longer wavelengths rather than enhanced by shorter wavelengths of light.

It has been suggested that the growth response to wavelength in broilers is age-dependent, with green light inducing faster initial growth and blue light producing larger body weight gains after 2 weeks (8). A transfer from green to blue light at 10 d of age was reported to have resulted in growth that was faster than constant blue, green or white light, although body weight at 46 d was not significantly different from that of birds reared throughout on green light (5). However, the blue (480 nm), green (560 nm) and white light in these experiments was at a common irradiance of 0.1 W/m² at bird height, and so the birds' perception of light intensity would have been different for each light source, and responses to wavelength may have been confounded with those to illuminance.

Feed intake and conversion

When broilers were offered feed illuminated by green, yellow, orange or red light in the same room the highest intakes were from the feeders lit by green light (17). However, there were no differences when a single colour was used within a room, indicating that the observed behaviour was a preference for green light rather than a direct response to this part of the spectrum. This supports the unequivocal conclusion that wavelength has no effect on feed intake (1-10). Thus, the similarity of feed intake across the visual spectrum, but adverse effect of long wavelengths of light on growth, results in broilers and growing turkeys converting feed more efficiently under blue and green light than under red.

• Male reproduction •

Assessments of the effect of wavelength on sexual development are complicated because gonadal growth is strongly influenced by light information received by both the retina and the hypothalamus and, in addition to the differences in retinal sensitivity, penetrability to and sensitivity at hypothalamic photoreceptors also varies with wavelength. Whichever criterion is used to measure sexual maturation in males (weight of testes, pituitary gland or comb, or seminiferous tubule diameter), the greatest stimulation is achieved by exposure to white or red light. This has been demonstrated in fowl (1, 2, 6), turkeys (18), quail (19) and Mallard drakes (20, 21). The only contradictory report is for 32-d old broiler males, which showed no influence of red, green, blue or white light on testicular weight, but had significantly higher plasma testosterone concentrations under blue than green, with intermediate concentrations for white and red (7). However, it is likely that these birds were too young to be photosexually responsive (see page 25).

The strong influence of red and white light (which includes red light) has also been demonstrated in sexually mature domestic cockerels (22). An abrupt change from a stimulatory 14-h photoperiod to a non-stimulatory 6-h resulted in reductions in testicular weight being progressively larger under blue, green, red and white photoperiods, respectively. Additionally, lower testes weight and inhibitions of spermatogenesis and gonadotrophin release were recorded at 18 weeks in cockerels reared under red or white 6-h photoperiods, compared with 14-h photoperiods, but no differences were observed between the two daylength groups when lit with blue or green light (23). The more stimulatory effect of longer wavelengths of light was also demonstrated when, following implantation of blue or yellow-orange radio-luminous beads into the brain of immature male and female quail, rapid gonadal growth occurred in birds implanted with the yellow-orange beads (24).

• Female reproduction •

In many investigations, the failure to equate irradiance or illuminance, the use of broad-spectrum commercial lamps, the introduction of natural light or the provision of complex photoperiodic treatments has resulted in equivocal or incorrectly assessed effects of wavelength on female sexual maturation.

Sexual maturation

Despite these difficulties, it seems that, as in males, illumination by red or white (which includes red) light stimulates earlier gonadal development than blue or green light in force-moulted laying hens (25), in turkeys (26) and in quail (18). In two trials where pullets were given a 7 or 8-h decrease in photoperiod during rearing (23, 27), the later maturity of pullets exposed to white or red light, compared with blue or green, led the authors to the erroneous conclusion that shorter wavelengths stimulated reproductive development. However, the converse is the case, because the later maturity of the red and white treatments, following a reduction in daylength at 14 weeks, indicates greater, not lesser, photoresponsiveness. Thus, these findings support rather than contradict the general conclusions.

Where there were no significant differences in sexual maturity associated with coloured lighting, mitigating circumstances suggested by investigators included an overriding effect of high light intensities (43 to 724 lux) over wavelength in turkeys (28), easier intracranial tissue penetration in first-year pullets compared with post-moulted pullets (25), and extremely broad spectrum emissions from coloured lamps (29). In the latter report, the birds were housed in windowed pens, and this might have resulted in all groups receiving sufficient stimulation from the 10 to 12 h of daylight that came through the windows. When brown-egg laying hybrids were reared under white, green or red light, and transferred to white or red illumination at different ages between 15 and 20 weeks, a combination of green followed by red light advanced maturity compared with pullets maintained on white light (30). Sexual maturity occurred 11 d earlier when the transfer from green to red light was made at 15 weeks and 6 d earlier following the transfer at 20 weeks. The smaller advance at the older age suggests that the pullets may have been influenced by a change from a non-stimulatory to a stimulatory wavelength, and are analogous with responses to daylength, where changes are more stimulatory than constant photoperiods and where the amplitude of the response to an increase in daylength decreases as the pullet approaches spontaneous maturity (see page 25).

Egg production

In contrast to its retarding effect on gonadal development, short-wavelength illumination has generally had a minimal influence on the rate of egg production in laying hens (31-34) and breeding turkeys (28, 35). Consistently heavier egg weights were observed from the beginning of the laying cycle for blue and green illuminated domestic hens than for birds exposed to red or white light (33). However, it is likely that the shorter wavelengths of light delayed sexual maturation, and that it was this, and not the colour differences, that increased egg weight.

The addition of 8 h of ultraviolet (principally UV-A) radiation to an 8-h photoperiod of white light was reported to have no effect on mean oviposition time (36). This is not surprising, because photosexual responses are mainly initiated by encephalic light reception, and the shorter the wavelength the less efficiently it penetrates the skull and brain tissues to the hypothalamus (see page 10).

• Behaviour, preference, stress and mortality •

Activity and aggression

Despite there being few reports of behavioural responses to coloured light, those that are available support the commonly held belief that blue light has a calming effect on birds.

Broilers are less active and spend more time sitting passively and dozing under blue or green light than under red or white light In contrast, when illuminated with red light they are more aggressive and do more wing stretching and floor pecking than birds illuminated with white, green or blue (3, 4). Male turkeys have also been observed to be more docile, less active, to show less sexually related activity and to have fewer social interactions under 5 lux of blue light than under red or white (9). However, these differences were not seen when the same colour comparisons were made at 86 lux. Low illuminance and short wavelength light are both associated with a delay in sexual maturity, and so it is not surprising that a combination of both factors would result in less sexual activity and a reduced incidence of the behavioural problems associated with it. However, when long wavelengths of light and bright stimulatory photoperiods occur singly or in combination, male turkeys commence sexual maturation towards the end of the rearing period and sexually related aggressive activity increases.

Significant behavioural differences have been observed in growing pullets reared under green, red or white fluorescent lighting with red light completely preventing cannibalism, whilst white and green light resulted in incidences of 30 and 41% cannibalism, respectively (31). This contrasts with the experience of poultry farmers that laying birds are, in common with the birds in the broiler studies (3, 4), more aggressive under red light than other colours, and that there is more cannibalistic behaviour under incandescent than under fluorescent lighting. It is ironical that laying hens are frequently transferred from white to red light to curb outbreaks of cannibalism. However, if red lamps are used to replace white lamps at the same wattage, there will be a reduction in illuminance as well as a change in colour, and the lower light intensity will likely be the factor that controls the aggression. It will, therefore, be more economical to change to a lower wattage white lamp than to the same wattage red. Data in Table 6.1 show that an incandescent light source produces a higher proportion of red light than any other white light source, with more than double the amount present in natural light, and it may be that it is the red component of incandescent light that triggers the aggression.

Coloured light preference

It is possible that poultry are able to identify a lighting environment in which interactions with their pen mates are most harmonious. When broilers, which had previously been reared under blue, green, red or white light were given a choice of illumination, birds reared under white, red or green light preferred the blue pen, whereas the 'blue' birds showed a preference for the green pen, a novel, but still short wavelength environment (3). Laying hens show a preference for fluorescent over incandescent white lighting (37), and growing turkeys spend more time in fluorescent than incandescent light (38) but have a preference for fluorescent supplemented with UV-A radiation (39). This indicates that poultry either have an aversion to long wavelengths of light or a preference for shorter wavelengths. Alternatively, they may prefer the lighting environment that is closest to natural light and which most fully satisfies their visual requirement.

Stress

There were no behavioural indications of stress when mature turkey hens were exposed to 16-h photoperiods of blue, green, red or white illumination from commercial lamps. Additionally, at the mid points of the photoperiod and the scotoperiod, plasma erythrocyte, leukocyte and corticosterone concentrations and red and white blood cell counts were not significantly different for any of the lighting treatments (40). Whilst these findings suggest that turkeys do not find blue, green or red light stressful, it is interesting that the birds exposed to the red light had the lowest proportion of heterophils and the narrowest heterophil: lymphocyte ratio of the four groups, and this is often an indicator of stress. However, it was concluded that the absence of any differences in

Table 6.1 *Typical proportional radiant fluxes within the wavelength bands of human colour sensations and avian UV sensitivity (350-780 nm) for various types of light source.*

Light source	UV-A (350-380)	Violet (380-435)	Blue (435-500)	Green (500-565)	Yellow (565-600)	Orange (600-630)	Red (630-780)
Sunlight	6.7	9.9	18.1	17.1	8.9	7.3	32.0
Incandescent:							
GLS lamp	0.2	1.0	3.6	8.7	7.6	8.4	70.5
Candle	0.0	0.3	1.6	5.3	5.7	7.0	80.0
Luminescent:							
Warm-white fluorescent lamp	3.3	10.3	8.5	24.7	14.1	29.7	9.3
Cool-white fluorescent lamp	0.2	5.4	17.1	40.8	13.3	20.7	2.4
High-pressure sodium lamp	0.4	1.2	4.5	6.8	52.1	18.0	16.9

egg production and the haematological and hormonal evidence suggested that the turkeys found none of the light sources stressful In contrast, turkeys illuminated with red and green light had significantly higher anti-sheep red blood cells (SRBC) titres 14 d (but not 7 or 21 d) after an intravenous injection of SRBC (40). This indicates that wavelength can affect the turkey's humoral immune response, and that this probably involves IgG rather than IgM.

Mortality

Surprisingly, despite birds frequently perceiving light from the various monochromatic light sources as a different illuminance, wavelength does not generally have a significant effect on mortality rates in broilers (7) or turkeys (9, 10). Similar conclusions may be drawn for broilers (14-16) and turkeys (12) when they are exposed to commercial lamps of different colours. However, mortality has been reported to be higher in growing pullets reared under red compared with white incandescent light (32). In contrast, broiler breeders illuminated with green light had better liveability between 10 and 40 weeks than under white light (41).

• References •

1. Foss, D.C., Donovan, G.A. & Arnold, E.L. (1967) The influence of narrow bands of light energy on growth, testis weight, pituitary weight and gonadotrophin production of male chickens. *Poultry Science* 46, 1258.
2. Johnson, A.L., Foss, D.C. & Carew, L.B. (1982) Effect of selected light treatments on pineal weight and lipid content in the cockerel (Gallus domesticus). *Poultry Science* 61, 128-134.
3. Prayitno, D.S., Phillips, C.J.C. & Omed, H. (1997) The effects of color of lighting on the behaviour and production of meat chickens. *Poultry Science* 76, 452-457.
4. Prayitno, D.S., Phillips, C.J.C. & Stokes, D.K. (1997) The effects of color and intensity of lighting on behavior and leg disorders in broiler chickens. *Poultry Science* 76, 1674-1681.
5. Rozenboim, I., Biran, I., Chaiseha, Y., Yahav, S., Rosenstrauch, A., Sklan, D. & Halevy, O. (2004) The effect of a green and blue monochromatic light combination on broiler growth and development. *Poultry Science* 83, 842-845.
6. Foss, D.C., Carew, L.B. Jnr. & Arnold, E.L. (1972) Physiological development of cockerels as influenced by selected wavelengths of environmental light. *Poultry Science* 51, 1922-1927.
7. Wabeck, C.J. & Skoglund, W.C. (1974) Influence of radiant energy from fluorescent light sources on growth, mortality and feed conversion of broilers. *Poultry Science* 53, 2055-2059.
8. Rozenboim, I., Biran, I., Uni, Z., Robinzon, B. & Halevy, O. (1999) The involvement of monochromatic light in growth, development and endocrine parameters of broilers. *Poultry Science* 78, 135-138.
9. Gill, D.J. & Leighton, A.T. Jnr. (1984) Effect of light environment and population density on growth performance of male turkeys. *Poultry Science* 63, 1314-1321.
10. Levenick, C.K. & Leighton, A.T. Jnr. (1988) Effects of photoperiod and filtered light on growth, reproduction and behaviour of turkeys *(Meleagris gallopavo)*. 1. Growth performance of two lines of males and females. *Poultry Science* 67, 1505-1513.
11. Lewis, P.D., Perry, G.C. & Sherwin, C.M. (1998) Effect of photoperiod and light intensity on the performance of intact male turkeys. *Animal Science* 66, 759-767.
12. Kondra, P.A. (1961) The effect of colored light on growth and feed efficiency of chicks and poults. *Poultry Science* 40, 268-269.
13. Cherry, P. & Barwick, M.W. (1962) The effect of light on broiler growth. I. Light intensity and colour. *British Poultry Science* 3, 31-39.

14. Proudfoot, F.G. & Sefton, A.E. (1978) Feed texture and light treatment effects on the performance of chicken broilers. *Poultry Science* 57, 408-416.
15. Wathes, C.M., Spechter, H.H. & Bray, T.S. (1982) The effects of light illuminance and wavelength on the growth of broiler chickens. *Journal of Agricultural Science, Cambridge* 98, 195-201.
16. Proudfoot, F.G. & Hulan, H.W. (1987) Interrelationships among lighting, ambient temperature, dietary energy and broiler chicken performance. *Poultry Science* 66, 1744-1749.
17. Smith, L.T. & Phillips, R.E. (1959) Influence of colored neon lights on feed consumption in poults. *Poultry Science* 38, 1248.
18. Gill, D.J. & Leighton, A.T. Jnr. (1988) Effects of light environment and population density on growth performance of male turkeys: Physiological changes. *Poultry Science* 67, 1518-1524.
19. Woodard, A.E., Moore, J.A. & Wilson, W.O. (1969) Effect of wavelength of light on growth and reproduction in Japanese quail. *Poultry Science* 48, 118-123.
20. Benoit, J., Walter, F.X. & Assenmacher, I. (1950) Nouvelles recherches relatives à l'action de lumières de différentes longueurs d'onde sur la gonadostimulation du canard male impubère. *Chronicles of the Royal Society of Biology* 144, 1206.
21. Benoit, J., Walter, F.X. & Assenmacher, I. (1950) Contribution à l'étude du réflexe optohypophysaire. Gonadostimulation chez le canard soumis à des radiations lumineuses de diverses longueurs d'onde. *Journal of Physiology* 42, 537-541.
22. Harrison, P.C., Latshaw, J.D., Casey, J.M. & McGinnis, J. (1970) Influence of decreased length of different spectral photoperiods on testis development of domestic fowl. *Journal of Reproduction and Fertility* 22, 269-275.
23. Casey, J.M., Harrison, P.C., Latshaw, J.D. & McGinnis, J. (1969) Effects of photoperiod and colored lights on the sexual maturity of domestic fowl. *Poultry Science* 48, 1794.
24. Homma, K., Ohta, M. & Sakakibara, Y. (1980) Surface and deep photoreceptors in photoperiodism in birds, in: *Biological rhythms in birds: Neural and endocrine aspects* (Eds. Tanabe, Y., Tanaka, K. & Ookawa, T.), pp. 149-156. Tokyo, Japan Scientific Societies Press/ Berlin, Springer Verlag.
25. Pyrzak, R., Snapir, N., Goodman, G., Arnon, E. & Perek, M. (1986) The influence of light quality on initiation of egg laying by hens. *Poultry Science* 65, 190-193.
26. Scott, H.M. & Payne, L.F. (1937) Light in relation to the experimental modification of the breeding season of turkeys. *Poultry Science* 16, 90-96.
27. Harrison, P., McGinnis, J., Schumaier, G. & Lauber, J. (1969) Sexual maturity and subsequent reproductive performance of white leghorn chickens subjected to different parts of the light spectrum. *Poultry Science* 48, 878-883.
28. Pyrzak, R. & Siopes, T.D. (1986) Effect of light quality on egg production of caged turkey hens. *Poultry Science* 65, 199-200.
29. Carson, J.R., Junnila, W.A. & Bacon, B.F. (1958) Sexual maturity and productivity in the chicken as affected by the quality of illumination during the growing period. *Poultry Science* 37, 102-112.
30. Foss, D.D. & White, J.L. (1983) Early sexual maturity of brown-egg pullets cage-grown in narrow-band light with high nutrient density diets. *Poultry Science* 62, 1424.
31. Schumaier, G., Harrison, P.C. & McGinnis, J. (1968) Effect of colored fluorescent light on growth, cannibalism and subsequent egg production of single comb white leghorn pullets. *Poultry Science* 47, 1599-1602.
32. Wells, R.G. (1971) A comparison of red and white light and high and low dietary protein regimes for growing pullets. *British Poultry Science* 12, 313-325.
33. Pyrzak, R., Snapir, N., Goodman, G. & Perek, M. (1984) The influence of light quality on egg production and egg quality of the domestic hen. *Poultry Science* 63 (Suppl), 30.
34. Rozenboim, I., Zilberman, E. & Gvaryahu, G. (1998) New monochromatic light source for laying hens. *Poultry Science* 77, 1695-1698.

35. Jones, T.E., Hughes, B.L., Thurston, R.T., Hess, R.A. & Froman, D.P. (1982) The effect of red and white light during the prebreeder and breeder periods on egg production and feed consumption in Large White turkeys. *Poultry Science* 61, 1930-1932.
36. Lewis, P.D., Perry, G.C. & Morris, T.R. (2000) Ultraviolet radiation and laying pullets. *British Poultry Science* 41, 131-135.
37. Widowski, T.M., Keeling, L.J. & Duncan, I.J.H. (1992) The preferences of hens for compact fluorescent over incandescent lighting. *Canadian Journal of Animal Science* 72, 203-211.
38. Sherwin, C.M. (1999) Domestic turkeys are not averse to compact fluorescent lighting. *Applied Animal Behaviour Science* 64, 47-55.
39. Moinard, C. & Sherwin, C.M. (1999) Turkeys prefer fluorescent light with supplementary ultraviolet radiation. *Applied Animal Behaviour Science* 64, 261-267.
40. Scott, R.P. & Siopes, T.D. (1994) Light color: effect on blood cells, immune function and stress status in turkey hens. *Comparative Biochemical Physiology* 108A, 161-168.
41. Cave, N.A. (1990) Effects of feeding level during pullet-layer transition and of pretransition lighting on performance of broiler breeders. *Poultry Science* 69, 1141-1146.

7 PATHOLOGICAL EFFECTS OF LIGHTING

This chapter describes the deleterious effects that very long photoperiods, continuous illumination, continuous darkness, low illuminance and coloured light can have upon the integrity of the eye. It also reports the adverse effects of long daylengths on adrenal function and immunosuppression in poultry, and the incidence of ovarian carcinoma in turkeys. A glossary of ophthalmic terms and ocular anatomical names and a diagram of the anatomy of the domestic fowl's eye are included for reference (Figure 7.1).

• Glossary •

Ametropia is the inability to focus images on the retina due to an imperfection in refractive function. It can occur as myopia or hyperopia.

Anterior chamber is the space at the front of the eye between the cornea and the iris/lens that contains aqueous humour fluid.

Buphthalmos is abnormal enlargement of the eye, also termed buphthalmia.

Choroid is the vascular membrane between the sclera and the retina.

Ciliary body is a vascular part of the eye that connects the iris with the choroids.

Cornea is the transparent membrane covering the exposed front part of the eye.

Dioptre is a measure of refractive power in a lens, and is the reciprocal of its focal length (m).

Emmetropia is the condition of perfect vision, and its development is termed emmetropisation.

Fovea is a depression in the retina that contains only cones and is the area where images have the sharpest definition.

Glaucoma is an increase in intra-ocular pressure that results in imperfect vision.

Hyperopia is the condition when images are focused behind the retina, and is synonymous with long-sightedness.

Iris is the coloured muscular diaphragm at the front of the eye that adjusts the amount of light entering the lens.

Lens is the biconvex transparent structure behind the iris that focuses images on the retina.

Myopia is the condition when images are focused in front of the retina, and is synonymous with short-sightedness.

Pecten is a comb-like vascular structure that projects from the retina. It is peculiar to birds and reptiles, and is thought to provide the retina with oxygen.

Retina is the light-sensitive membrane that forms the inner lining of the vitreous chamber. Images focused on it by the lens are transmitted as neurochemical signals to the brain.

Sclera is the white fibrous membrane that forms the outer covering of the eye, and is continuous with the cornea.

Sclerotic ring is a circle of bony plates that provides protection to the front of the eye. It is necessary as compensation for the loss of strength that results from the avian eye not being spherical.

Vitreous or posterior chamber is the space at the back of the eye between the lens and the retina that contains vitreous humour fluid.

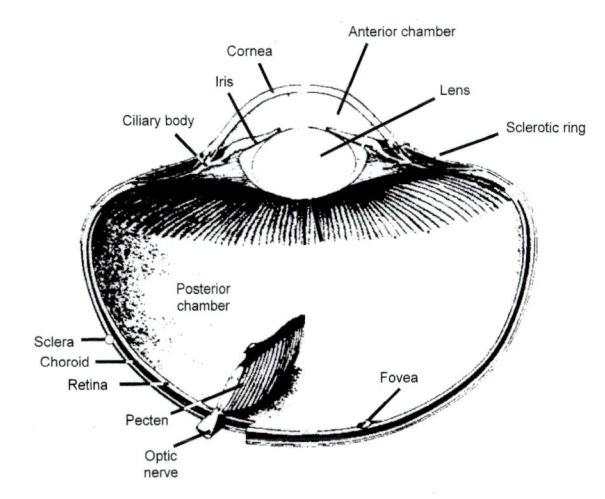

Figure 7.1 *Transverse section of the domestic fowl eye.*

Chapter 7 Pathological effects of lighting

• Lighting extremes for chicken and turkeys •

Constant illumination
Eye abnormalities: domestic fowl

Enlargement of the eye in birds exposed to continuous illumination (LL) was first reported in 6-week old chickens illuminated with incandescent light (1). The increase in eye diameter was thought to have been caused primarily by an accumulation of fluid. Buphthalmia and corneal flattening were subsequently recorded in young chickens transferred to fluorescent LL conditions at 1040 lux (2). Increased intra-ocular pressure in these birds occurred within 7 d of LL, but the condition lasted for only 14 d before the plastic nature of the eye allowed a gradual stretching and enlargement of the globe which reduced the pressure. Buphthalmia has been observed in young chicken kept on ≥ 22-h photoperiods as well as LL (3, 4), resulting in shallow anterior chambers and increased hyperopia (Figure 7.2, 5).

Continuous illumination has also resulted in deepened vitreous chambers, corneal thickening, lenticular thinning, cataracts, damage to the retina (loss of oil droplets), pigment epithelia and choroids in pullets at 17 weeks of age (Table 7.1, 5). However, the corneal flattening, increased vitreous chamber depth and hyperopia were apparent from 10 d. It is normal for the cornea to flatten in birds during emmetropisation, but the corneal radius increases at a steeper rate in LL birds so that by 11 weeks it can be almost double that of 14-h controls (Figure 7.3, 3).

Chicks hatch hyperopic, with a refractive error of about 5 dioptres, but during emmetropisation there is a natural reduction to reach a nadir of ≤ 3 dioptres by 6 weeks of age. However, chickens exposed to continuous illumination become progressively more hyperopic and reach an asymptote of 17-18 dioptres by 9 to 10 weeks of age (Figure 7.4, 5).

The pathological effects of continuous or near-continuous artificial lighting on the structure and function of the eye have been observed under both incandescent (1, 2) and fluorescent (3, 5) illumination, and over a range of intensities from 25 (1) to 5000 (4) lux.

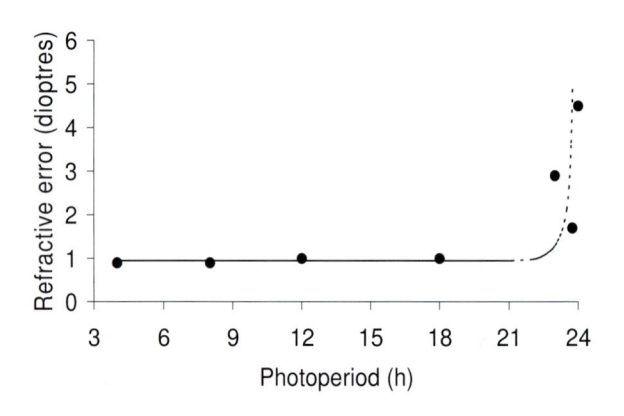

Figure 7.2 *Refractive error in 2-week-old chickens following exposure to 4, 8, 12, 18, 23, 23.75-h photoperiods or LL illumination from a fluorescent light source (5).*

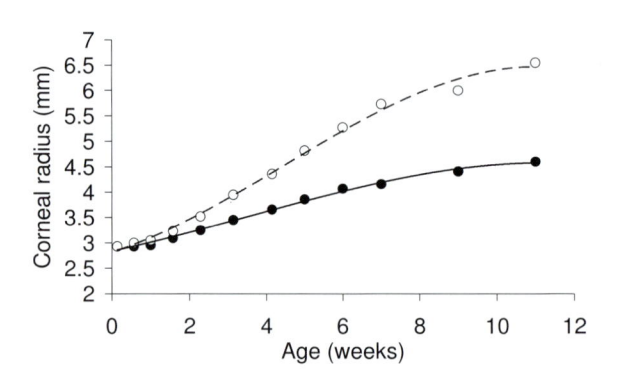

Figure 7.3 *Changes in corneal radius with age for chickens given a 14-h photoperiod (●) or continuous illumination (○) from a fluorescent lamp at 700 lux (3).*

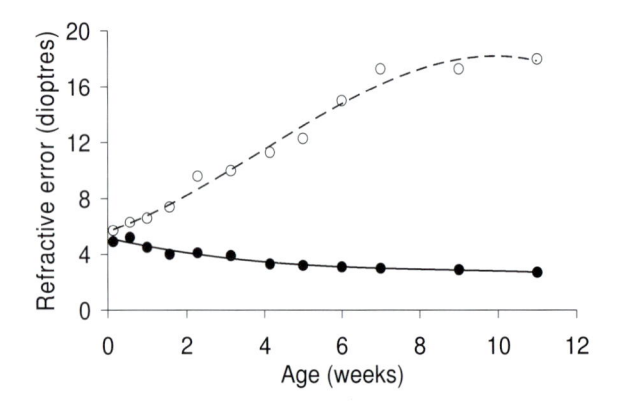

Figure 7.4 *Temporal changes in refractive error for chickens exposed to 14-h photoperiods (●) or continuous illumination from a fluorescent lamp (○) at 700 lux (5).*

Table 7.1 *Ocular dimensions (mm) of eyes from 17-week-old pullets reared on 14L:10D or continuous illumination (LL) (5).*

Measurement	LL	14L:10D
Cornea diameter	9.29[b]	10.30[a]
Lens diameter	7.53	7.45
Eye diameter	19.32[a]	17.54[b]
Cornea radius	8.25[a]	5.18[b]
Ant. chamber depth	0.84[b]	2.37[a]
Post. chamber depth	10.45[a]	8.46[b]
Lens thickness	3.26[b]	3.81[a]
Axial length	14.62	14.57

Means with different superscripts are significantly different at P<0.05

Buphthalmia was not observed in broilers that had been continuously illuminated for 8 weeks with 12 h of natural light followed by 12 h of incandescent light (6). This may indicate that ocular abnormalities are only induced by continuous illumination when the light lacks something that is peculiar to natural light. One possibility is ultraviolet radiation, because birds use it for forming visual images and as a visual cue for controlling behaviour. Whereas 7 weeks of continuous exposure to supplemental UV radiation did not induce buphthalmia or affect the vision of broiler chickens given 23 h of white incandescent lighting, it did cause a roughening of the corneal surfaces (7).

When genetically blind (rc/rc) chicks which had normal papillary reflexes but eyes 18% heavier than normally-sighted controls were exposed to constant light, there was no further enlargement of the eye, even though anterior chambers became more shallow and corneas flatter than sighted controls (8).

Prolonged exposure to LL conditions (up to 21 months) has been reported to result in retinal detachment, cupping of the optic disc, cataracts and ingress of the eye interior with spongy bone and pro-cartilage in laying hens (9).

Eye abnormalities: growing turkeys

Turkeys exposed to 23-h photoperiods or continuous illumination show similar eye abnormalities to those observed in LL broiler and laying birds (10, 11, 12). Additionally, turkeys develop elongated eyelids (Figure 7.5,

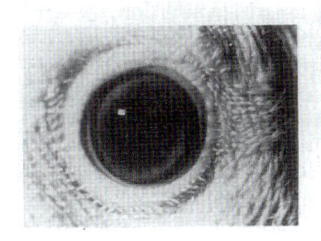

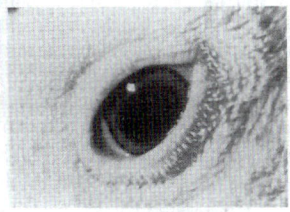

Figure 7.5 *Effect of exposure to 12-h photoperiods (left) or continuous illumination (right) from mixed incandescent (<20 lux) and indirect natural (<220 lux) lighting on eye-lid shape in turkey poults at 8 weeks (12)*

12), have increased sclera exposure and become blind (10, 11). In one investigation, the severity of the changes in eyelid shape, sclera exposure and cornea curvature, which had been noticed in continuously illuminated birds (incandescent light at 30 or 2000 lux) from as early as 7 d and present in 70% of birds by 6 weeks, was markedly reduced after the birds were transferred to natural lighting. However, the abnormalities developed again when the birds were returned to continuous lighting (10).

Elongation of the eyelid seems to disrupt normal optical and visual processes, and increases in the anterior to posterior diameter make the birds short-sighted and reluctant to move away from familiar surroundings (10, 12).

Immunosuppression

The provision of continuous incandescent illumination at 45 lux has been shown to result in immunosuppression in immature cockerels (13). This was characterised by reduced antibody production, compared with 12L:12D controls, in response to secondary immunisations with sheep red blood cells (SRBC) at 13 and 18 weeks of age. There was also a slower development of hypersensitivity in response to injections of phytohaemagglutinin (PHA-P) at 12 weeks and concanavalin (Con A) at 16 weeks (13).

Stress

Serological tests and adrenal weights have been equivocal in their demonstration of the effects of continuous illumination on the stress status of chickens and turkeys. Whereas constant light conditions resulted in increases in relative adrenal mass, cholesterol concentration and plasma free-fatty-acid concentration in chickens

Table 7.2 *Relative adrenal mass and cholesterol concentration in domestic pullets at 3 weeks (14) and turkeys at 8 weeks (15) exposed to 12L:12D or continuous illumination, and plasma free fatty acid (FFA) and corticosterone concentrations for chickens.*

	12L:12D	LL
Chickens		
Adrenal mass (mg/kg)	144±7	174±4
Adrenal cholesterol (mg/g)	16.9±0.6	19.2±0.7
FFA (µM/l)	142±8	203±13
Corticostreone (ng/ml)	8.1±0.8	6.7±0.6
Turkeys		
Adrenal mass (mg/kg)	95	80
Corticosterone (ng/ml)	17.2	16.3

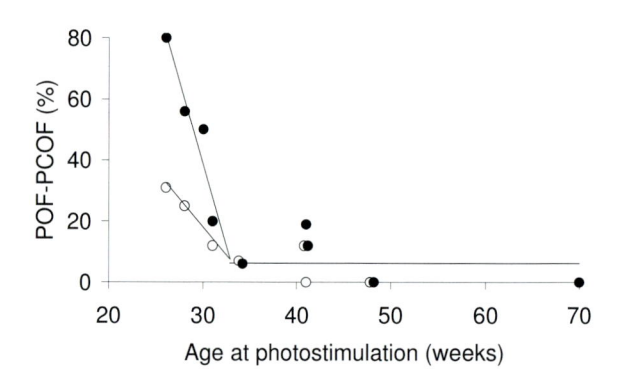

Figure 7.6 *The effect of age at photostimulation on the incidence of polyovarian follicle-polycystic ovarian follicle syndrome in turkeys transferred from 6L:16D to 14L:10D (○) or LL (●) (18).*

at 3 weeks, compared with 12L:12D-controls (14), relative adrenal mass was significantly smaller for 8-week-old turkeys (15) (Table 7.2).

Plasma corticosterone concentration evidence is also unclear, with a significant reduction reported for LL chicken (14) but no significant effect in turkeys (15). Responses have also differed between trials in the same report, with LL broiler chicken having significantly higher plasma corticoid concentrations than either a symmetrical or asymmetrical intermittent lighting regimen in one trial but no significant difference in a second trial (16).

Polyovarian Follicle and Polycystic Ovarian Follicle Syndrome in turkeys

Recently, a condition termed Polyovarian Follicle and Polycystic Ovarian Follicle Syndrome (POF-PCOF) was described in breeding turkeys that had been transferred from 6-h photoperiods to continuous illumination at 30 weeks (17). The condition was characterised by hens that ceased to ovulate within 2 to 3 weeks of becoming sexually mature, but still recruited follicles into the ovarian hierarchy, some of which continued to grow and become cystic. Typically, birds with POF-PCOF had lower plasma luteinizing hormone (LH) but higher progesterone (P_4) concentrations than normal, and some birds retained a shelled egg in the oviduct for days or weeks. It has been suggested that the condition is caused by a constantly high plasma P_4 concentration that blocks pre-ovulatory LH surges, while normal concentrations of follicle stimulating hormone and oestradiol continue to stimulate recruitment (17). The incidence of POF-PCOF is significantly higher in turkeys transferred to LL than in those given 14L:10D, especially when the birds have been photostimulated at, or before, 32 weeks of age (Figure 7.6, 18).

Continuous darkness
Eye abnormalities

Broiler chickens given 1 to 2 weeks of conventional lighting prior to 4 months of continuous darkness (DD) developed eye abnormalities that were similar to those seen in chickens exposed to LL within 7 d of the light being withdrawn (19, 20). As is the case in LL birds, hyperopia develops as a consequence of corneal flattening. Initially, there is an increase in intra-ocular pressure (IOP), but normal IOP returns within 6 weeks despite continuing eye enlargement (19). This transient rise in IOP, which has also been observed in continuously lit birds, is most likely the result of the vitreous chamber enlarging more quickly than the development of plasticity in the eye membranes (2). Additionally, DD is associated with lethargic behaviour, with birds not shying away from others or foreign objects, and being less active than 12L:12D controls (19, 20).

Thinning of the choroid and retina, as manifest by darkened patches in the peripheral retina and non-pigmented white bands in the retina perpendicular to the pecten, have only been observed in DD conditions. However, this pathology is thought to be a consequence of slow development caused by a reduced blood supply to the choroid, rather than to any stretching of the retina (19).

Adrenal function
Transferring broiler males at 6 weeks of age from a 14L:10D lighting regimen to DD conditions was reported to have no effect on adrenal weights at 11 weeks compared with fluorescently illuminated 14-h controls (21).

General
Arrhythmic melatonin release
The inducement of ocular abnormalities in poultry kept in complete darkness as well as continuous illumination suggests that it is not light *per se* that causes the pathology but a consequence of the absence of some circadian rhythm (5). The diurnal pattern of melatonin release is disrupted by both LL and DD conditions, and so arrhythmic melatonin release is a likely indicator for the development of eye abnormalities. Support for this suggestion is provided by an investigation in which chicks were exposed to LL conditions between 3 and 24 d, but with one eye, both eyes, the head (and thus the pineal gland) or all three organs covered with opaque devices for 12 h daily. The findings indicated that the provision of a diurnal light-dark cycle to any or all of the bird's melatonin-

Table 7.3. *Effect of covering one eye, both eyes, and or the head for 12 h daily in continuously illuminated (LL) chicks (22).*

Treatment: area covered in LL	Refractive error (dioptres)	Anterior chamber depth (mm)	Vitreous chamber depth (mm)
12L:12D	3.81	1.61	5.99
None	12.67	0.93	6.72
Head	6.50	1.21	6.33
One eye	6.25	1.50	6.16
Both eyes	5.81	1.53	6.03
Head & eyes	4.75	1.49	5.84

synthesising organs will protect it from LL-induced eye abnormalities, because a normal circadian rhythm of melatonin release is maintained (Table 7.3, 22).

Lethargy
Constant LL and DD conditions have both been observed to induce lethargy, and it has been suggested that this is caused by a generalised metabolic stress (19).

Genotypic variations
Genotypes vary in the degree to which they respond to extremes of lighting, even within the same breed (White Leghorn) (3, 23).

• Illuminance and wavelength •

Illuminance
Although eye enlargement observed in egg-type chickens exposed to continuous illumination at 5000 lux was reported to be less evident than at 150 lux, both groups suffered some damage to the cornea (24).

The provision of 23-h photoperiods at 1 lux has been shown to cause eye abnormalities and adrenal hypertrophy in 2-week-old turkeys in contrast to 11, 110 or 220 lux (25). The ocular pathology included flattened corneas, heavier eye weights and increased transverse eye diameter. The 1-lux illuminance was also associated with reduced activity and increased huddling.

Coloured light
Eye abnormalities
Eye enlargement, characterized by exophthalmia (protrusion of the eye) rather than buphthalmia, developed in young chickens exposed to 14 h of very dim blue (maximum 0.046 µW/cm² at 450 nm) light (26, 27, 28). However, whereas buphthalmia develops within 3-4 weeks when chicken are exposed to LL or DD conditions, eye enlargement was not apparent in these birds until about 7 weeks of age. Unfortunately, it is not possible to conclude whether wavelength or low illuminance caused the eye enlargement in these experiments because the two factors were confounded.

Stress

Coloured light does not appear to impose a consistent stress on poultry. Adrenal weights in 11-week-old cockerels given 5 weeks of far-red (peak of 750 nm), near-red (650 nm), green (545 nm) or blue (450nm) light were similar to constant white-light controls (21). Serological studies also showed that the effects of colour are equivocal in turkeys. The heterophil:lymphocyte ratio in turkey poults exposed to 16 h of red light (peak 615 nm) was significantly lower than in birds illuminated with white or blue (453 nm) light, with green light (548 nm) intermediate. Plasma corticosterone concentra-tions in the middle of both the light and dark periods were similar for all four treatment groups (30). Light in this trial was provided at a constant photon output, but the various colours would have been perceived to be of a different illuminance because of the inverse relationship between the energy of a photon and its wavelength (page 1). However, it is unlikely that illuminance affected the responses to wave-length since the white-light intensity was moderately high (43 lux). Observations of behaviour also revealed nothing indicative of stress which, together with similar white and red blood cell counts, suggests that wavelength has minimal effect on the stress level of turkeys. In contrast, colour did have some effect on the immune-response, though this was dependent on how long after SRBC inoculation the blood sample was taken. Anti-SRBC titres were similar for all wavelengths 7 d after inoculation, significantly higher for poults exposed to red or green light than to blue or white after 14 d, but similar again for all groups after 21 d. These findings suggest that the influence of wavelength on the primary immune-response involves immunoglobulin G. Responses to a secondary inoculation of SRBC were similar for all lighting groups (30). The light-induced suppression of melatonin synthesis is not equal at all wavelengths, at least in hamsters (31), and so it is possible that the effect of colour on the immune-response is achieved through a modification of the circadian rhythm of melatonin release.

Growing pullets developed hypothyroidism and had reduced oxygen consumption after 7 weeks of exposure to 14 h photoperiods of very dim blue light (26).

UV-A radiation
Eye abnormalities

Continuous exposure to supplementary UV-A radiation has resulted in broilers given a conventional 23L:1D lighting regimen developing a roughening of the surface of the cornea and a thickening of the corneal epithelium (7). However, there was no adverse effect on vision or any sign of eye enlargement.

Stress

In contrast to wavelengths of visible light, supplemental UV-A radiation has been reported to result in lower basal plasma corticosterone concentrations (and presumably a lower stress status) in chickens illuminated with white light from a halogen lamp than in controls given white light only (32). The birds exposed to UV-A also tended to conduct more exploratory behaviour, though this did not lead to an increase in the number of aggressive encounters or to feather pecking.

• Ovarian carcinoma in turkeys •

Turkey hens held on long days (16 h) can develop spontaneous ovarian adenocarcinoma, especially when they are kept for more than one laying cycle. Such tumours were found to have completely regressed within 4.4±0.5 weeks when hens were changed from 16 to 8-h photoperiods, but had regenerated within 5-6 weeks of the birds being transferred back to 16 h (33). It is known that tumours develop in animals exposed to long days as a consequence of the decreased synthesis of melatonin by the pineal gland (34), and work with turkeys has shown that a daily injection of 50 µg melatonin significantly delays tumour regeneration in hens exposed to 16-h photoperiods (33).

Chapter 7 Pathological effects of lighting

• References •

1. Jensen, L.S. & Matson, W.E. (1957) Enlargement of avian eye by subjecting chicks to continuous incandescent illumination. *Science* 125, 741.
2. Whitley, R.D., Albert, R.A., Brewer, R.N., McDaniel, G.R., Pidgeon, G.L. & Mora, E.C. (1984) Photoinduced buphthalmic avian eyes. I. Continuous fluorescent light. *Poultry Science* 63, 1537-1542.
3. Stone, R.A., Lin, T., Desai, D. & Capehart, C. (1995) Photoperiod, early post-natal eye growth, and visual deprivation. *Vision Research* 35, 1195-1202.
4. Oishi, T. & Murakami, N. (1985) Effects of duration and intensity of illumination on several parameters of the chick eye. *Comparative Biochemistry and Physiology A* 81, 319-323.
5. Li, T., Troilo, D., Glasser, A. & Howland, H.C. (1995) Constant light produces severe corneal flattening and hyperopia in chickens. *Vision Research* 35, 1203-1209.
6. Shutze, J.V., Jensen, L.S., Carver, J.S. & Matson, W.E. (1960) Influence of various lighting regimes on the performance of broiler chickens. *Washington Agricultural Experimental Station Technical Bulletin* 36, pp. 1-11.
7. Barnett, K.C. & Laursen-Jones, A.P. (1976) The effect of continuous ultraviolet irradiation on broiler chicks. *British Poultry Science* 17, 175-177.
8. Lauber, J.K. & Oishi, T. (1989) Ocular responses of genetically blind chicks to the light environment and to lid suture. *Current Eye Research* 8, 757-764.
9. Lauber, J.K. & McGinnis, J. (1966) Eye lesions in domestic fowl reared under continuous light. *Vision Research* 6, 619-626.
10. Barnett, K.C., Ashton, W.L., Holford, G., Macpherson, I. & Simm, P.D. Chorioretinitis and buphthalmos in turkeys. *The Veterinary Record*, 620.-627.
11. Ashton, W.L., Pattison, M. & Barnett, K.C. (1973) Light-induced eye abnormalities in turkeys and the turkey blindness syndrome. *Research in Veterinary Science* 14, 42-46.
12. Davis, G.S., Siopes, T.D., Peiffer, R.L. & Cook, C. (1986) Morphologic changes induced by photoperiod in eyes of turkey poults. *American Journal of Veterinary Research* 47, 953-955.
13. Kirby, J.D. & Froman, D.P. (1991) Research Note: Evaluation of humoral and delayed hypersensitivity responses in cockerels reared under constant light or a twelve hour: twelve hour dark photoperiod. *Poultry Science* 70, 2375-2378.
14. Freeman, B.M., Manning, A.C.C. & Flack, I.H. (1981) Photoperiod and its effect on the responses of the immature fowl to stressors. *Comparative Biochemistry and Physiology* 68A, 411-416.
15. Davis, G.S. & Siopes, T.D. (1985) The effect of light duration on turkey poult performance and adrenal function. *Poultry Science* 64, 995-1001.
16. Buckland, R.B., Bernon, D.E. & Goldrosen, A. (1976) Effect of four lighting regimes on broiler performance, leg abnormalities and plasma corticoid levels. *Poultry Science* 55, 1072-1076.
17. Liu, H. K., Long, D.W. & Bacon, W.L. (2001) Concentration change patterns of luteinizing hormone and progesterone and distribution of hierarchal follicles in normal and arrested laying turkey hens. *Poultry Science* 80, 1509-1518.
18. Bacon, W.L. & Liu, H.K. (2003) Influence of photoperiod and age of photostimulation on the incidence of polycystic ovarian follicle syndrome in turkey breeder hens. *Poultry Science* 82, 1985-1989.
19. Jenkins, R.L., Ivey, W.D., McDaniel, G.R. & Albert, R.A. (1979) A darkness induced eye abnormality in the domestic chicken. *Poultry Science* 58, 55-59.
20. Whitley, R.D., Albert, R.A., McDaniel, G.R., Brewer, R.N., Mora, E.C. & Henderson, R.A. (1985) Photoinduced buphthalmic avian eyes. II. Continuous darkness. *Poultry Science* 64, 1869-1874.
21. Foss, D.C., Carew, L.B., Jnr. & Arnold, E.L. (1972) Physiological development of cockerels as influenced by selected wavelengths of environmental light. *Poultry Science* 51, 1922-1972.

22. Li, T. & Howland, H.C. (2003) The effects of constant and diurnal illumination of the pineal gland and the eyes on ocular growth in chicks. *Investigative Ophthalmology and Visual Science* 44, 3692-3697.
23. Troilo, D., Glasser, A, Li, T & Howland, H. (1992) Different strains of chick have different eye growth responses to visual deprivation. *Investigative Ophthalmology and Visual Science* 33, 711.
24. Oishi, T. (1980) Light- and dark-induced eye lesions in birds, in: *Biological rhythms in birds* (Eds. Y. Tanabe., K. Tanaka. & T. Ookawa). pp. 249-256, Berlin, Springer-Verlag.
25. Siopes, T.D., Timmons, M.B., Baughman, G.R. & Parkhurst, C.R. (1984) The effects of light intensity on turkey poult performance, eye morphology, and adrenal weight. *Poultry Science* 63, 904-909.
26. Harrison, P.C., Bercovitz, A.B. & Leary, G.A. (1968) Development of eye enlargement of domestic fowl subjected to low intensity light. *International Journal of Biometeorology* 12, 351-358.
27. Bercovitz, A.B., Harrison, P.C. & Leary, G.A. (1972) Light induced alterations in growth pattern of the avian eye. *Vision Research* 12, 1253-1259.
28. Harrison, P.C. & McGinnis, J. (1967) Light induced exophthalmos in the domestic fowl. *Proceedings of the Society for Experimental Biology and Medicine* 126, 308-312.
29. Chiu, P.S.L., Lauber, J.K. & Kinnear, A. (1975) Dimensional and physiological lesions in the chick eye as influenced by the light environment. *Proceedings of the Society for Experimental Biology and Medicine* 148, 1223-1228.
30. Scott, R.P. & Siopes, T.D. (1994) Light color: effects on blood cells, immune function and stress status in turkey hens. *Comparative Biochemistry and Physiology* 108, 161-168.
31. Brainard, G.C., Richardson, B.A., King, T.S. & Reiter, R.J. (1984) The influence of different light spectra on the suppression of pineal melatonin content in the Syrian hamster. *Brain Research* 294, 333-341.
32. Maddocks, S.A., Cuthill, I.C., Goldsmith, A.R. & Sherwin, C.M. (2001) Behavioural and physiological effects of absence of violet wavelengths for domestic chicks. *Animal Behaviour* 62, 1013-1019.
33. Moore, C.B. & Siopes, T.D. (2004) Spontaneous ovarian adenocarcinoma in the domestic turkey breeder hen (Meleagris gallopavo): Effects of photoperiod and melatonin. *Neuroendocrinology Letters* 25, 94-101.
34. Lapin, V. & Ebels, I. (1981) The role of the pineal gland in neuroendocrine control mechanisms of neoplastic growth. *Journal of Neural Transmission* 52, 275-282.

8 LIGHT SOURCE

In recent years, many poultry producers have changed from tungsten-filament lamps (incandescent) to more energy efficient, longer lasting light sources. These include low-pressure mercury (fluorescent) and high-pressure sodium vapour discharge lamps that have 4 to 5 times the luminous efficiency, and 10 to 30 times the life of incandescent lamps (Table 8.1). Discharge lamps have the advantage that they can be made to produce light with different spectral characteristics, but fluorescent light is sometimes emitted at a frequency that may be perceived by birds as discontinuous. This chapter describes the characteristics of various light sources commonly used in poultry houses, and the responses to them.

• Light generation and characteristics of light sources •

Light generation

The main difference between the various types of lamp is the method by which the light is produced. Additionally there are differences in luminous efficiency, initial cost, running cost and working life (Table 8.1).

Incandescence

The oldest and still most common way of generating light is incandescence, a method that involves heating a solid body to the point at which light is emitted. It can be achieved using various types of body, including the glowing carbon particles in the flame of a candle or paraffin lamp, the mantle of a gas or pressurised-paraffin lamp or the filament of an incandescent, more correctly a general lighting service (GLS), lamp. In a GLS lamp, an electric current is passed through a thin tungsten wire of high resistance so as to heat it to about 2800 K.

The main disadvantage of an incandescent lamp is its inefficiency; less than 10% of the energy produced is emitted as visible light, with the remainder being heat (Figure 8.1). Advantages for incandescence include its non-dependence on electricity, cheapness, ease of dimming and the continuous spectrum of the light produced.

Luminescence

Luminescence is light produced by excitation of electrons of an atom, most commonly obtained by sending an electric current through a gas, such as neon or mercury vapour. Unlike the continuous spectrum of incandescently produced light, luminescent light is produced in a discontinuous spectrum with spikes occurring at various wavelengths (Figure 8.1).

Fluorescent lamps use a low-pressure mercury discharge that initially produces a high proportion of radiation in the UV range, but then converts this short-wave radiation into longer wavelength visible light through a process called fluorescence. This is achieved by coating the inside of the discharge tube with a fluorescent powder called a phosphor. Many phosphors are available, and they can be used singly or in combination to produce radiation with different wavelength peaks, thus enabling the manufacture of various fluorescent lamps such as warm-white, cool-white and daylight.

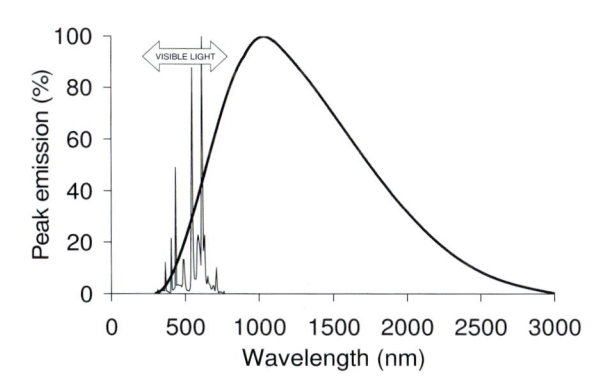

Figure 8.1 *Typical luminous fluxes (% of peak emission at 5 nm intervals) for an incandescent GLS (continuous line) and a warm-white fluorescent lamp (spiked line).*

Table 8.1 *Typical characteristics for commonly used light sources.*

Light source	Power range (W)	Colour temperature (K)	Warm-up time	Luminous efficiency (lumens/W)	Average life (h)
GLS (Incandescent)	15-1,000	2850	Negligible	10-18	1,000
Standard fluorescent tubes	15-125	2900 [1]	1-2 s	45-72	15,000
Compact fluorescent bulbs	5-55	4000 [2]	1-2 s	35-70	10,000
High-pressure sodium vapour	50-1,000	2000	3.5-5 min	52-105	28,500
Low-pressure sodium vapour	18-180	1700	8-10 min	90-142	16,000
Metal halide	100-2,000	4500	3.5-5 min	62-102	14,000

[1] Warm-white (standard or compact)
[2] Cool-white (standard or compact)

Table 8.2 *Typical spectral composition between 300 and 780 nm for various fluorescent lamps.*

Wavelength (colour sensation)	Warm-white fluorescent (%)	Cool-white Fluorescent (%)	Daylight fluorescent (%)
300-400 (UV-A)	0.4	0.3	0.2
400-435 (violet)	4.9	5.3	6.2
435-500 (blue)	10.0	17.1	25.9
500-565 (green)	37.4	40.8	41.5
565-600 (yellow)	16.8	13.3	10.5
600-630 (orange)	27.3	20.7	13.9
630-780 (red)	3.3	2.4	1.8

There is also a lamp designed specifically for lighting birds that emits ultraviolet (UV-A) in addition to white light, thus fully satisfying the bird's visual needs (1).

Low-pressure sodium lamps use neon as a starting gas to initiate a discharge of sodium vapour, producing yellow light predominantly in two very narrow bands at 589.0 and 589.6 nm (resonance lines D_1 and D_2). High-pressure sodium lamps use a xenon starting gas and a mercury buffer gas, absorb the D_1 and D_2 line, and broaden the other emission bands to give almost continuous spectrum white light.

Lamp characteristics
GLS incandescent lamps
GLS lamps are compact, have a low initial cost, perfect colour rendering (i.e. the ability to reveal colours), simplicity of installation and do not need a warming-up period. However, their running costs are comparatively high and their average working life is relatively short (about 1,000 h). GLS lamps are easily dimmed with voltage reduction equipment, but in so doing the spectral composition of the light is altered. When voltage is reduced, the filament temperature falls and peak energy output occurs at a longer wavelength, resulting in warmer, redder light. For example, at 100% voltage the light from a GLS lamp has a colour temperature of about 2850 K and peak emission around 1000 nm, but at 80% voltage the colour temperature drops to about 2600 K, peak emission moves to 1100 nm, and light output reduces to just under half that at 100% (Figure 8.2, 2). If lower illuminance is required without a change in light colour, it is necessary to either use a lower wattage lamp or provide some form of physical screening. Reductions in voltage and light output are not proportional; below 50% voltage the light output is negligible (Figure 8.2, 2).

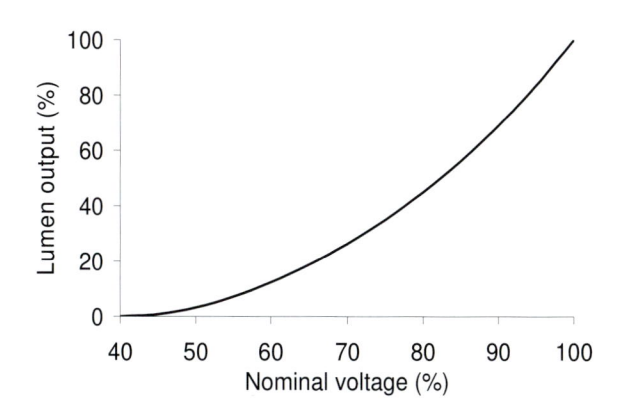

Figure 8.2 *Effect of reducing voltage on light output of a GLS lamp (2).*

Fluorescent lamps

Fluorescent lamps produce light much more efficiently than GLS, but they cannot be connected directly to a mains electricity supply without a device to limit the electrical current flow (ballast), and a 'starter' to switch the preheating current and provide a high enough voltage to vaporise the mercury. As a result, the initial costs of lamps and fittings are higher than GLS. A 1-2 s warm-up period is required and conventional fluorescent tubes can only be dimmed using special equipment. Compact fluorescent lamps have the advantage of being able to directly replace GLS bulbs in the same fittings, be they bayonet or screw, and are produced in lower wattages than conventional fluorescent tubes. However, as with fluorescent tubes, conventional voltage reduction equipment cannot be used to dim a standard compact lamp, and starting problems can sometimes occur at low temperatures.

Sodium vapour lamps

High-pressure sodium vapour lamps, despite having a much longer life and greater efficiency than fluorescent lamps (Table 8.1), are not often used in poultry houses because individual units are too bright and a 3-5 min warm-up period is required. Other forms of luminescent lighting, such as low-pressure sodium vapour, mercury vapour and metal halide, though potentially more efficient and longer lasting, are even less suitable for illuminating poultry because of extended warm-up periods, high minimum output ratings, poor colour rendering and/or undesirable spectral characteristics.

Coloured lamps

When GLS lamps are used to produce coloured light, the primary light produced during incandescence is still white, but the output is coloured because pigments on the glass filter out other wavelengths. However, fluorescent sources can produce coloured light, because the availability of diverse fluorescent powders, and mixtures thereof, permits the production of light with almost any desired colour rendering characteristic (Table 8.2).

Perception of fluorescent light flicker
Flicker perception

Fluorescent light is discontinuous by physical standards, but is perceived by humans as continuous because the highest flicker rate detectable by the human eye, called the critical fusion frequency (CFF), is between 40 and 60 Hz, depending on illuminance (Table 8.3, 3).

Flicker sensitivity is caused by two interacting processes operating within the retina; a chemical diffusion that creates a time delay when photons are converted into nerve impulses and an inhibitory feedback involving horizontal and amacrine cell connections (4). Generally, domestic fowl (and presumably other poultry) have a flicker sensitivity that is markedly lower than humans, that is they only detect flicker at a higher modulation. However, at high frequencies, humans and fowl sensitivities are similar; both have maximum sensitivity at 10-15 Hz (Figure 8.3, 3). The CFF for fowl is similar to humans up to a luminance of 500 cd/m² (equal to an illuminance of 125 lux at 2 m from the light source), but is slightly higher than humans at 1000 cd/m² (Table 8.3, 3). The highest reported CFF in laying hens is 105 Hz which was observed in monochromatic blue light at a quantum flux of 784 photons/s (5). The similarity between the flicker threshold data for humans and laying hens suggest that they have temporal visual systems that are controlled by analogous mechanisms (3).

Fluorescent lighting used in poultry houses is usually modulated at low frequencies, for example, 100 Hz (Europe) or 120 Hz (America), and so under European frequencies and at high intensities poultry may see the direct light from fluorescent sources (driven by 50 Hz alternating current, AC) as a flicker (5). This could occur

Table 8.3 *Critical flicker fusion frequencies for humans and laying hens at various levels of luminance (3).*

Luminance (cd/m²)	Critical fusion frequency (Hz)	
	Human	Hen
10	40.8	39.2
100	50.4	54.0
200	53.3	54.0
500	58.2	57.4
1000	57.4	71.5

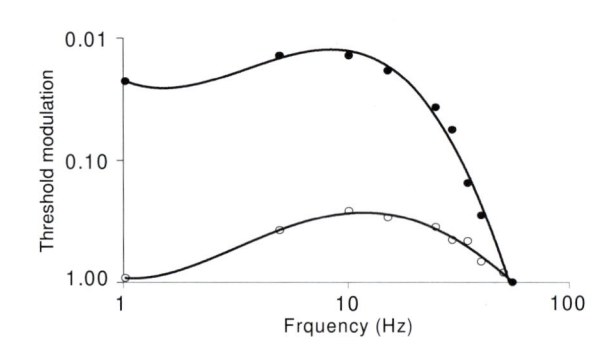

Figure 8.3 *Flicker sensitivity (threshold modulation) for humans (●) and laying hens (○) at a luminance of 200 cd/m² (3).*

for laying hens in cages located directly opposite a fluorescent lamp. Light reflected from metal objects and the corneal surface of other birds' eyes is also likely to appear as discontinuous at high light intensities, but the diffused reflections from the walls and ceilings of poultry houses probably appear continuous.

Asymmetrical discharges

Fluorescent tubes have two discharges in each AC cycle that are in opposite directions, for example, 50 Hz AC electricity has 50 discharges per second in each direction. However, old fluorescent lights can become asymmetrical, *i.e.,* they have discharges in one direction that are of a different magnitude from those in the other direction. This asymmetry produces an impression of flicker for humans, and is more than likely to produce flickering surroundings for poultry. During fast movement, such as pecking, birds illuminated by asymmetrical lamps will see other birds as a series of stroboscopic images. This raises important welfare questions and may imply that fluorescent lighting should be regularly checked for signs of asymmetry.

Fluorescent preference

Notwithstanding that light from a fluorescent source powered by a standard low-frequency ballast may be perceived as discontinuous by domestic poultry at some light intensities, laying hens do not appear to find it aversive; on the contrary they seem to find some components of this form of lighting attractive. When laying hens were allowed to move freely between rooms illuminated by either compact fluorescent or incandescent lamps that had been set to give 12 lux illuminance at floor level, the hens spent significantly more time in the fluorescent than in the incandescent light (6). All the main physical activities (feeding, drinking, nesting, preening, perching, sitting, standing and walking) were performed in both types of light, although disproportionately more preening was performed under fluorescent light. Many species of birds have UV-A reflective plumage (7), and so it is likely that birds prefer to preen under shorter wavelengths of light because their plumage is more readily visible than under redder light (see Table 6.1, page 99 for spectral outputs of fluorescent and incandescent lamps). When pullets were given a choice between low or high-frequency compact fluorescent lighting at 10-12 lux, there was no indication that the birds had a preference for either light source or that the sources had any influence upon their behaviour (8). These findings suggest that fluorescent lighting at an illuminance commonly used in commercial poultry houses, whether powered by a low or high-frequency ballast, does not compromise the welfare of the laying bird.

Fluorescent anti-rachitic properties

More than 70 years ago, it was discovered that a daily exposure of 3 min UV radiation was sufficient to prevent rickets in chicks given diets deficient in vitamin D (10, 11). Fluorescently illuminated 21-d chicks were shown to have tibiae with a 30% higher fat-free ash content than incandescently lit birds. Other fluorescently illuminated chicks had bone calcification that was similar to that of birds given food supplemented with 173.9 mg cod liver oil/100 g, which is equivalent to 200 AOAC units of vitamin D/kg (9).

A small proportion of fluorescent light is produced in the UV range (see Table 6.1, page 99), so this no doubt explains the beneficial effects of fluorescent light in birds kept indoors and given diets deficient in vitamin D.

However, the anti-rachitic property of fluorescent light is irrelevant for modern intensively housed poultry whose diets are routinely fortified with adequate vitamin D.

• Production responses to light source •

Broilers

In many studies of the response of broilers to light source, lamp type has been confounded with wavelength, illuminance or both. Nevertheless, it seems clear that neither growth nor feed conversion efficiency is significantly affected by light source, at least for birds grown to between 42 and 70 d of age (12-16), because light source did not interact with intensity in any of the trials. For example, similar body weights have been recorded under a wide range of light sources with illuminance ranging from 2 lux (15W incandescent bulbs) to 20 lux (40W fluorescent tubes).

Although light source also appears not to influence mortality, it may influence the prevalence of certain skeletal abnormalities. In a comparison of pink fluorescent and white incandescent illumination, more angular deformities and total leg abnormalities occurred in the incandescently lit birds, but a higher incidence of tibial dyschondroplasia was seen in the birds grown under fluorescent light. However, it should be noted that light source was confounded with colour in this trial (13).

Growing pullets

There is minimal evidence for the effect of light source on sexual maturation, and even then it is contradictory. Broiler breeders reared on varying intensities of fluorescent, incandescent or high-pressure sodium lighting were reported to have similar rates of sexual development (17), but egg-type pullets illuminated with incandescent lamps matured significantly later than others illuminated with warm-white fluorescent light (18). In both of these investigations, lamp type was confounded with illuminance, but, since light intensity in all treatments exceeded the 4-lux threshold for maximising sexual development (see Figure 5.2, page 81), the observed rates of gonadal development are unlikely to have been influenced by illuminance. The almost complete absence of any influence of light source on adult reproduction suggests that it is not involved in the photosexual response, and so it is likely that sexual development is also unaffected by the type of illumination.

Laying hens and broiler breeders

In general, egg laying performance in commercial layers (19-23) and broiler breeder hens (17, 24, 25) is unaffected by light source. However, there are some reports that light source has affected egg production at different stages of the laying period. In egg-type hens, rate of lay was noted to be sporadically higher under fluorescent than under incandescent lighting (20), to be superior after 58 weeks in broiler breeders illuminated with incandescent as opposed to cool-white fluorescent lamps (25), and to be higher under natural light supplemented with incandescent light than incandescent illumination alone (17). Shell quality, fertility and hatchability, at least in broiler breeders, seem to be unaffected by sources of illumination (24).

Oviposition time and entrainment

Similar mean oviposition times and entrainments to 'dawn' and 'dusk' signals have been reported for hens subjected to bright:dim ahemeral cycles created by various combinations and light intensities of compact fluorescent, incandescent and fluorescent tube light sources (22).

Feed intake

Although feed intake has not generally been influenced by light source, a trend was observed in one study for feed intake to be higher in laying hens illuminated with fluorescent than incandescent light (21). Activity has also been found to be higher under fluorescent than under incandescent lighting (26), and so it is possible that extra feed may have been used to satisfy an increased demand for energy.

Mortality and behaviour
Mortality during the laying year has, without exception, been unaffected by light source (19, 22-24), and agonistic behaviour was similar in laying hens exposed to fluorescent or incandescent light, even though the light was supplied at different intensities (20).

Breeding turkeys
Sexual maturation
Interestingly, responses to alternative sources of artificial illumination for breeding turkeys have been studied for more than 60 years. In 1945, it was reported that turkeys matured up to 5 weeks earlier when birds were given natural lighting supplemented with light from gasoline, natural gas or incandescent lamps to give continuous illumination than birds exposed to natural light alone (27). Somewhat surprisingly, birds given extra light from a kerosene lamp matured at a similar time to the naturally illuminated birds, but the authors reported that the kerosene lamp was frequently blown out in windy weather because of a cracked glass! Clearly the responses here were to the increase in photoperiod and not to the light sources. More recently, age at first egg was reported as similar for turkeys reared under fluorescent and incandescent lighting (28, 29).

Egg production and female fertility
Although there are some data to suggest that light source during the rearing period may have an effect upon subsequent egg production, the evidence is equivocal, and it is likely that the rearing light source has no effect on rate of lay. Egg production was similar in two trials in which the birds had been reared under fluorescent, high-pressure sodium vapour or incandescent illumination (30, 31). In a third trial, hens that had been reared under high-pressure sodium vapour had a superior rate of lay to others given incandescent light at intensities of 5 and 30 lux, though it was inferior at 180 lux (32). Egg production during a typical 20-week laying cycle is generally unaffected by the source of illumination in the laying facilities (28-34). However, light source has sometimes differentially affected rate of lay within the laying cycle. Poorer egg production has occurred during the first 10 weeks of the laying cycle under incandescent compared with fluorescent or high-pressure sodium vapour lighting (30), but better egg production was recorded in the second half of the production period under incandescent than under fluorescent illumination (28, 35).

In general, light source had no significant effect upon egg weight, shell quality, fertility, hatchability, food intake or body weight gain during the laying cycle, though in one study heavier egg weights were recorded under high-pressure sodium vapour lamps than under incandescent lighting (31).

Male fertility
Although increased semen volumes and numbers of sperm per ejaculate were recorded in males given 14 h of incandescent light, when compared with birds given the same daylength given as a mixture of natural and incandescent light, there were no effects on fertility and hatchability (36). In a second investigation, sperm volume and semen production was maintained between 52 and 68 weeks of age in toms illuminated from a daylight fluorescent source but reduced progressively to a complete cessation in birds given cool-white fluorescent lighting (37). The birds under daylight fluorescent light also had significantly higher plasma testosterone levels at 68 weeks and normal spermatogenesis in most seminiferous tubules, while sperm production was at various stages of degeneration in the birds under cool-white fluorescent tubes. Testicular weights were similar for both groups. However, it is difficult to believe that the light sources could have been responsible for the reported differences in fertility since they have very similar spectral compositions (see Table 8.2, page 112).

Growing turkeys

Although light source seems to have no influence on the growth of females to 16 (38) or 33 weeks of age (30), significant but contradictory differences have been recorded at intermediate ages. In one study, body weights at 10 and 14 weeks were significantly lower in one trial, but higher in another, when birds were illuminated with incandescent rather than fluorescent or low-pressure sodium lighting (38). In contrast, the body weight of parent females between 14 and 22 weeks was consistently lower under incandescent than under high-pressure sodium or fluorescent illumination (30). Feed conversion by females has been inconsistently affected by light source, being significantly less efficient under low-pressure sodium lighting than under fluorescent or incandescent in one trial but unaffected by light source in another (38). Growth and feed conversion efficiency to 24 weeks in males was similar under incandescent, two types of fluorescent and low-pressure sodium lighting (39).

It seems likely that, as is the case for broiler chicken, light source has no effect on growth or feed conversion efficiency in either male or female turkeys.

Mortality, skeletal integrity and behaviour

Mortality has generally not been affected by light source, although there is an isolated report of losses in males being significantly higher under fluorescent than under low-pressure sodium or incandescent lighting (29).

The incidence of leg abnormalities and live-bird quality have also been unaffected by the type of lighting.

In most studies of the effects of light source on performance in growing turkey, the birds have been beak-trimmed, toe-clipped and de-snooded. Thus, any undesirable behaviour that might have been induced by the type of lighting was less likely to be detected. However, the behavioural responses to lighting that have been observed have been complex, with inconsistent differences in agonistic behaviour and the incidence of feather pecking. In one study, males illuminated with warm-white fluorescent or low-pressure sodium light performed significantly more 'social pecking' than under incandescent light, but, in another trial, there was significantly less pecking under daylight fluorescent than under incandescent, warm-white fluorescent or low-pressure sodium illumination (29). Differences in spectral distribution among the light sources used in these trials will have contributed to the observed inconsistencies in behavioural response because wavelength has been shown to have an effect on turkey behaviour (see page 98). When male turkeys have been left intact, the incidences of tail and wing injuries under fluorescent have been less than under incandescent light (40).

Geese

A study of the responses of geese to supplementation of natural light with cool-white fluorescent, daylight fluorescent or incandescent light to create a 16-h daylength indicated that significantly more eggs were laid over a 26-week laying cycle under cool-white fluorescent lighting than under daylight fluorescent or incandescent illumination (41).

• General conclusions •

In many trials, light source has been confounded with illuminance and/or spectral composition, and this may have contributed to some of the anomalies that have been reported. However, evidence from studies involving growing pullets, laying hens, broilers, growing turkeys, breeding turkeys and geese suggests that, irrespective of the light's spectral composition or illuminance, growth and reproductive performance are similar under modern energy efficient and conventional incandescent sources of illumination. A striking feature of the evidence is its inconsistency. There are no cases where 'significant' results have been corroborated by data from other studies.

Egg production

Differences observed in the egg production of laying hens or turkey breeders have only been recorded intermittently or in one part of the laying cycle. Light source appears to have no effect upon egg quality or any hatchability parameter, and, almost without exception, no influence on egg weight.

Growth and leg integrity

Light source has no adverse effect on body weight gain or food conversion efficiency in meat-type chickens or turkeys. However, in one broiler trial, fluorescent lighting reduced the overall incidence of leg problems but increased the numbers of birds suffering with tibial dyschondroplasia when compared with incandescent lighting. Light source has no effect on mortality in either broilers or growing turkeys.

Welfare concerns

Concerns that fluorescent light may be perceived as discontinuous and, therefore, detrimental to bird welfare seem to be misplaced. There have been no reported differences in the incidence of agonistic behaviour or, with the exception of one report in male turkeys, mortality in any species. Indeed, when laying hens were given a choice, they spent more time in fluorescent than in incandescent lighting, and turkey males have been reported to practise more 'social or non-aggressive' pecking under fluorescent than under incandescent lighting.

UV exposure limits

Recent studies have indicated that the provision of UV-A radiation (probably light to birds) in addition to white light may improve the visual environment of poultry kept in closed housing (43). It would, however, be prudent to check the exposure limit values for protection from ultraviolet radiation before asking anyone to work under a light source that emits ultraviolet radiation. The International Commission on Non-Ionizing Radiation Protection (ICNIRP) recommends an effective radiant exposure limit for unprotected skin of 30 J m^{-2} ultraviolet radiation, which is 0.001 W m^{-2} effective irradiance in an 8-h period (42). The corresponding unweighted radiant limit for retinal exposure is 10 kJ m^{-2}.

• References •

1. Arcadia bird-lamps. www.arcadia-uk.com
2. Pritchard, D.C. (1995) Lamps, in: *Lighting*, pp. 54-81. Longman, Harlow.
3. Jarvis, J.R., Taylor, N.R., Prescott, N.B., Meeks, I. & Wathes, C.M. (2002) Measuring and modelling the photopic flicker sensitivity of the chicken (*Gallus g. domesticus*) *Vision Research* 42, 99-106.
4. Kelly, D.H. (1971) Theory of flicker and transient responses 1: uniform fields. Journal of the Optical Society of America 61, 632-640.
5. Nuboer, J.F.W., Coemans, M.A.J.M. & Vos, J.J. (1992) Artificial lighting in poultry houses: do hens perceive the modulation of fluorescent lamps as flicker? *British Poultry Science* 33, 123-133.
6. Widowski, T.M., Keeling, L.J. & Duncan, I.J.H. (1992) The preference of hens for compact fluorescent over incandescent lighting. *Canadian Journal of Animal Science* 72, 203-211.
7. Bennett, A.T.D. & Cuthill, I.C. (1994) Ultraviolet vision in birds: What is its function?. *Vision Research* 34, 1471-1478.
8. Widowski, T.M. & Duncan, I.J.H. (1996) Laying hens do not have a preference for high-frequency versus low-frequency compact fluorescent light sources. *Canadian Journal of Animal Science* 76, 177-181.
9. Willgeroth, G.B. & Fritz, J.C. (1944) Influence of incandescent and fluorescent lights on calcification in the chick. *Poultry Science* 23, 251-252.
10. Scott, H.T., Hart, E.B. & Halpin, J.G. (1929) Winter sunlight, ultra violet light, and glass substitutes in the prevention of rickets in growing chicks. *Poultry Science* 9, 65-76.
11. Mussehl, F.E. & Ackerson, C.W. (1931) Anti-rachitic value of S-1 Lamp radiation for chicks. *Poultry Science* 10, 68-70.

12. Andrews, D.K. & Zimmermann, N.G. (1990) A comparison of energy efficient broiler house lighting sources and photoperiods. *Poultry Science* 69, 1471-1480.
13. Hulan, H.W. & Proudfoot, F.G. (1987) Effects of light source, ambient temperature and dietary energy source on the general performance and incidence of leg abnormalities of roaster chickens. *Poultry Science* 66, 645-651.
14. Proudfoot, F.G. & Hulan, H.W. (1987) Interrelationships among lighting, ambient temperature, dietary energy and broiler chick performance. *Poultry Science* 66, 1744-1749.
15. Scheideler, S.E. (1990) Effect of various light sources on broiler performance and efficiency of production under commercial conditions. *Poultry Science* 69, 1030-1033.
16. Zimmermann, N.G. (1988) Broiler performance when reared under various light sources. *Poultry Science* 67, 43-51.
17. Van Krey, H.P. & Weaver, W.D. Jnr. (1988) Effect of various supplemental light sources and light intensity on egg production by broiler breeder hens. *Archiv für Gelflügelkunde* 52, 221-226.
18. Siopes, T.D. & Wilson, W.O. (1980) Participation of the eyes in the photostimulation of chickens. *Poultry Science* 59, 1122-1125.
19. Carson, J.R., Bacon, B.F. & Junnila, W.A. (1957) Light quality and the stimulation of egg production. *Poultry Science* 36, 1108.
20. Fitzsimmons, R.C. & Newcombe, M. (1990) The effects of fluorescent light sources on the performance of White Leghorn hens. *Poultry Science*, 69, 1455-1460.
21. Hill, A.M., Charles, D.R., Spechter, H.H., Bailey, R.A. & Ballantyne, A.J. (1988) Effects of multiple environmental and nutritional factors in laying hens. *British Poultry Science* 29, 499-511.
22. Rose, S.P., Bell, M. & Michie, W. (1985) Comparison of artificial light sources and lighting programmes for laying hens on long ahemeral light cycles. *British Poultry Science* 26, 357-365.
23. Schumaier, G., Harrison, P.C. & McGinnis, J. (1968) Effect of coloured fluorescent light on growth, cannibalism and subsequent egg production of single comb White Leghorn pullets. *Poultry Science* 47, 1599-1602.
24. Colman, M.A. & Minear, L.R. (1981) A comparison of the effects of fluorescent versus incandescent lighting on the reproductive performance of two strains of broiler breeders. *Poultry Science* 60, 1642.
25. Ingram, D.D., Biron, T.R., Wilson, H.R. & Mather, F.B. (1987) Lighting of end of lay broiler breeders: Fluorescent versus incandescent. *Poultry Science* 66, 215-217.
26. Boshouwers, F.M.G. & Nicaise, E. (1993) Artificial light sources and their influence on physical activity and energy expenditure of laying hens. *British Poultry Science* 34, 11-19.
27. Milby, T.T. & Thompson, R.B. (1945) Sources of artificial light for turkey breeding females. *Poultry Science* 10, 68-70.
28. Siopes, T.D. (1984) The effect of high and low intensity cool-white fluorescent lighting on the reproductive performance of turkey breeder hens. *Poultry Science* 63, 920-926.
29. Siopes, T.D. (1984) The effect of full-spectrum fluorescent lighting on reproductive traits of caged turkey hens. *Poultry Science* 63, 1122-1128.
30. Felts, J.V., Leighton, A.T. Jnr., Denbow, D.M. & Hulet, R.M. (1990) Influence of light source on the growth and reproduction of large white turkeys. *Poultry Science* 69, 576-583.
31. Hulet, R.M., Denbow, D.M. & Leighton, A.T. Jnr. (1992) The effect of light source and intensity on turkey egg production. *Poultry Science* 71, 1277-1282.
32. El Halawani, M.E., Waibel, P.E. & Noll, S.L. (1981) Effects of prebreeder light sources and nutrition and breeder nutrition and nest types on Large White turkeys' reproductive performance. *Poultry Science* 60, 1664.
33. Cavalchini, L.G. (1976) Effects of fluorescent and incandescent light on egg production of turkey breeding. *Proceedings of 5th European Poultry Conference, Malta,* pp. 843-849.
34. El Halawani, M.E., Waibel, P.E. & Noll, S.L. (1980) Effects of light source, air cooling and nutrition on breeder hen turkeys' reproductive performance. *Proceedings of Virginia Turkey Days 1980,* pp. 4-8.
35. Payne, L.F & McDaniel G.R. (1958) Fluorescent lights for turkey breeders. *Poultry Science* 37, 722-726.
36. Jones, J.E., Hughes, B.L. & Elkin, R.G. (1977) Effect of light intensity and source on tom reproduction. *Poultry Science* 56, 1417-1420.

37. Snapir, N., Perek, M. & Pyrzak, R. (1970) The effect of artificial sunlight spectrum illumination on reproductive traits of turkey males. *Proceedings of 5th European Poultry Conference, Malta,* pp. 1281-1286.
38. Denbow, D.M., Leighton, A.T. & Hulet, R.M. (1990) Effect of light source and light intensity on growth performance and behaviour of female turkeys. *British Poultry Science* 31, 439-445.
39. Leighton, A.T., Hulet, R.M. & Denbow, D.M. (1989) Effect of light sources and light intensity on growth performance and behaviour of male turkeys. *British Poultry Science* 30, 563-574.
40. Moinard, C., Lewis, P.D., Perry, G.C. & Sherwin, C.M. (2001) The effects of light intensity and light source on injuries due to pecking of male turkeys *(Meleagris gallopavo). Animal Welfare* 10, 131-139.
41. Pyrzak, R.N., Snapir, N., Robinzon, B. & Goodman, G. (1984) The effect of supplementation of daylight with artificial light from various sources and at two intensities on the egg production of two lines of geese. *Poultry Science* 63, 1846-1850.
42. ICNIRP (2004) Guidelines on limits of exposure to ultraviolet radiation of wavelengths between 180 nm and 400 nm (Incoherent optical radiation). *Health Physics* 87, 171-186.
43. Lewis, P.D., Perry, G.C., Sherwin, C.M. & Moinard, C. (2000) Effect of ultraviolet radiation on the performance of intact male turkeys. *Poultry Science* 79, 850-855.

9 LIGHTING FOR GROWING PULLETS AND LAYING HENS

This chapter describes lighting programmes for growing pullets and laying hens that will optimise performance, including the appropriate daylength, light intensity, light colour and lamp type. In addition to conventional programmes, it describes lighting to enhance egg weight or shell quality, to stimulate appetite when growth is below target, to maximise feed conversion efficiency and to control undesirable behaviour.

• Aims of a lighting programme for growing pullets and laying hens •

- Stimulate feed intake and growth
- Influence the timing of sexual maturity
- Maximise egg numbers
- Optimise egg weight
- Influence time of egg-laying
- Control undesirable behaviour

• Lighting during the rearing period •

Age and body weight at first egg strongly influence egg numbers and egg weight, and so the principal aims of a lighting programme during the rearing period are to get the pullets to sexual maturity at the right age and at the correct body weight to ensure that they produce the egg package required for the market. Although the main factor controlling the timing of sexual maturation is the age at which the pullets are given an increase in daylength, the time taken to reach the base daylength and the base daylength itself also have some effect.

Daylength
Normal maturity in lightproof houses
In controlled environment facilities, it is usual to give continuous illumination for the first one or two days after pullets arrive from the hatchery to allow them to find feed and water, and then to reduce the daylength in steps to reach a base daylength of between 8 and 10 hours by 2 weeks of age. The use of a step-down, rather than an abrupt decrease to the base daylength is not essential, but many producers feel that any change in management is better made progressively rather than abruptly. There is minimal difference in the subsequent growth and sexual development between the two approaches. If a step-down is used, it is best to keep the time of 'lights-out' constant and to shorten the day by adjusting 'lights-on'. Chicks quickly learn to fill their crops and then settle for the night before the anticipated sunset. If 'lights-out' is moved forward they are caught unawares. On the other hand, they will readily get up and feed before the lights come on if the night is a long one and if they feel hungry.

It was standard practice for many years to rear pullets on 8-hour daylengths. However, there has been a recent tendency by some breeding companies to recommend 9 or 10-hour daylengths because modern hybrids naturally mature earlier than their ancestors, and the use of longer daylengths stimulates feed intake and makes it easier to achieve breeding company's body weight targets (which have generally been raised as genetic sexual maturity has advanced). There is nothing to be gained from using daylengths longer than 10 hours. This does not advance maturity but actually has a small

Table 9.1 *The effect of a transfer from various base daylengths to a new daylength at 16 weeks of age on the age (days) at 50% egg production for pullets that would normally reach 50% lay at 20 weeks of age if maintained on 8-hour daylengths (bold font indicates the earliest maturity).*

Base daylength (hours)	New daylength (hours)					
	9	10	11	12	14	16
8	135	132	129	128	126	126
9	*	131	128	127	125	125
10	*	*	128	126	**124**	**124**
11	*	*	*	128	125	125
12	*	*	*	*	127	126

delaying effect on sexual development (Figure 3.1, page 24). Additionally, longer daylengths during the rearing period limit the opportunity to increase the daylength when the pullets are photostimulated at the end of the rearing period. Table 9.1 shows that, irrespective of the new daylength, the earliest maturity will be achieved by rearing pullets on 10 hours. It also shows that transferring to a daylength that is longer than 14 hours does not further advance the age that a flock reaches 50% lay. However, a longer base daylength will increase feed intake and produce a heavier body weight at the end of the rearing period, with cumulative feed intake to 18 weeks increasing by about 100 g between 8 and 12 hours (see Figure 3.2, page 24) and body

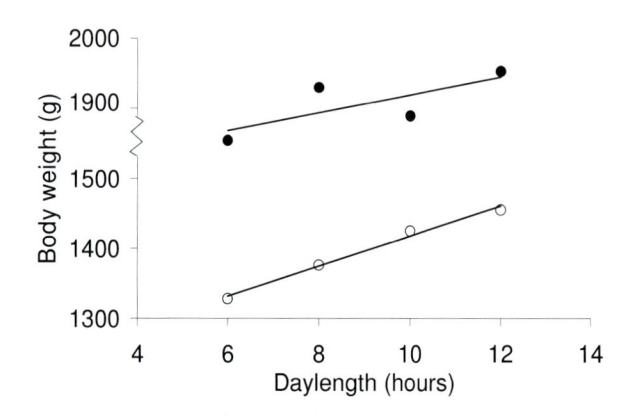

Figure 9.1 *Body weight at 18 weeks of age for brown-egg (●) and white-egg hybrids (○) reared on 6, 8, 10 or 12-hour daylengths (unpublished data from the University of Guelph).*

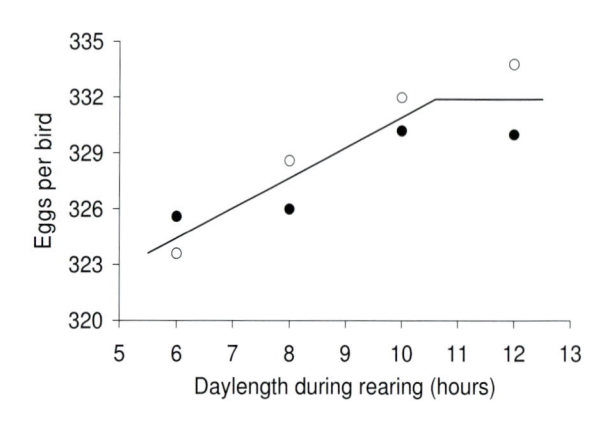

Figure 9.2 *Egg numbers to 70 weeks for brown-egg (●) and white-egg (○) hybrids reared to 18 weeks on 6, 8, 10 or 12-hours daylengths and transferred to 12.5 h followed by weekly increases of 30 minutes to 14 hours by 21 weeks (unpublished data from the University of Guelph).*

weight at 18 weeks increasing by 15 to 20 g (Figure 9.1) for each 1-hour extension of the base daylength. The use of daylengths that are longer than 10 hours is recommended by some breeding companies to maintain growth when temperatures are abnormally high. This increases feed intake, not only because it provides a longer feeding period, but also because it gives the pullets an opportunity to feed during the cooler part of the day.

The influence of any lighting programme on age at 50% lay persists throughout the laying year, resulting in differences in egg numbers at the end of a fixed laying period that reflect days lost or gained at the beginning of production. Unpublished data in Figure 9.2 from the University of Guelph for brown-egg and white-egg hybrids illustrate this effect.

Primary breeder recommended programmes
The major layer breeding companies vary in their recommendations on how to light their hybrids to achieve standard sexual maturity (that is when there is no special requirement for increased egg weight). Some guides recommend an 8-hour base daylength, whereas others advise 9 or 10 hours (Table 9.2). All advise the producer to take 6 or 7 weeks to reach the base photoperiod. Presumably this is intended to make attainment of body weight targets easier (see page 24), but it also has the effect of

Table 9.2 *Typical lighting programmes recommended by primary breeding companies to achieve standard sexual maturity in lightproof facilities.*

Age (days)	LOHMANN LSL (hours)	ISA BROWN (hours)	HYLINE BROWN (hours)	BOVANS GOLDLINE (hours)
0-2	24	22	21	23
3-7	16	20	21	21
8-14	14	18	19	19
15-21	13	16	17	17
22-28	12	15	15	15
29-35	11	13.5	13.5	13
36-42	10	12	12	11
43-49	9	11	10.5	9
50-56	8	10	9	9
57-63	8	10	9	9
64-70	8	10	9	9
71-77	8	10	9	9
78-84	8	10	9	9
85-91	8	10	9	9
92-98	8	10	9	9
99-105	8	12	9	9
106-112	8	12.5	9	9
113-119	8	13	10	10
120-126	8	13.5	11	11
127-133	9	14	12	12
134-140	10	14.5	13	13
141-147	11	15	14	14
148-154	12	15	14.5	14
155-161	13	15	15	14
162-168	14	15	15.25	14
169-175	14	15	15.5	14
176-182	14	15	15.75	14
183 onwards	14	15	16	14

Information in this table has been obtained from management guides produced by the following primary breeding companies (current in 2005).

Lohmann Tierzucht GmbH	ISA	Hendrix Poultry Breeders BV	Hy-Line International
Cuxhaven	Saint Brieuc	Boxmeer	Iowa,
Germany	France	The Netherlands	U.S.A.

delaying onset of lay by about a week compared with a programme reducing to the base daylength at 14 days.

There are variations in the size and age at which the first increment in daylength is given, with some suggesting a 1-hour increase and others 2 hours. These variations will have no effect on performance. The age at which the first increment is recommended varies from 14 to 18 weeks. Starting at 14 weeks will result in pullets maturing about a week earlier than at 18 weeks but egg weight being about 1 g lighter.

Normal maturity in non-lightproof houses

After the initial one or two days of continuous light, it is recommended that winter and spring-hatched pullets housed in inadequately light-proofed, naturally ventilated or open sided buildings are not reduced to 10 hours but to a daylength that is about one hour shorter than the longest natural daylength to be experienced during the rearing period. However, the light intensity inside a poultry house, whether naturally lit or poorly light-proofed, will be less than that outside, so it is probably acceptable to use sunrise to sunset times for clock-setting, even though the daylength perceived by the pullet in truly natural lighting conditions is not the period between sunrise and sunset but that between the beginning of civil morning twilight and the end of evening civil twilight (see Dawn and dusk section on pages 49-50 and data in Table 3.12). The avoidance of exposure to increasing daylengths during the rearing period will prevent precocious sexual maturity which leads to the production of small eggs and an increased risk of prolapse. In the 1960s, Trevor Morris promulgated the golden rule 'Never allow the light to increase during the rearing period' and this still applies today.

The base daylength will depend on the time of year at which the pullets are hatched and the latitude of the rearing facilities. The appropriate constant daylength to use during the rearing period is given in Table 9.3.

Delayed maturity in lightproof houses

The only reason for poultry farmers to retard maturity in egg-type hybrids is to improve egg weight. Although feed restriction can be used to delay maturity, as routinely occurs in broiler breeders, this does not improve egg weight because the pullet is no heavier when it becomes sexually mature. Egg weight is related to body weight at maturity and so, to improve egg weight, the bird must be heavier when it lays its first egg. This can only be achieved by modifying the lighting programme to actively delay maturity whilst still allowing the bird to grow.

In the early days of poultry lighting, a step-down programme called the 'Reading' pattern was shown to result in later sexual maturity than the 'King' programme that had been developed in America. This gave pullets a constant daylength from day old, which is closer to the type of lighting given to most modern pullets. Subsequently, it was demonstrated that the amount of delay in maturity was proportional to the duration of step-down lighting, with age at 50% lay occurring about one day later for each extra week taken to reach the base daylength. Thus if a 7-day delay is required to induce a 1-g increase in egg weight, the daylength must be reduced progressively from the initial 2 days of continuous illumination to reach the base daylength by 7 or 8 weeks of age. Similarly, a 2-g

Table 9.3 *Artificial daylengths (hours) for pullets kept at various latitudes and hatched at different times but exposed to a naturally increasing lighting pattern up to 18 weeks of age.*

Latitude	Nov (May)	Dec (June)	Jan (July)	Feb (Aug)
15°	12	12	13	13
20°	12	12	13	13
25°	12	12	13	13
30°	12	12	13	14
35°	12	13	14	14
40°	12	13	14	15
45°	12	13	15	15
50°	12	13	15	16
55°	12	14	16	16

Month of hatch in Northern (or Southern hemisphere)

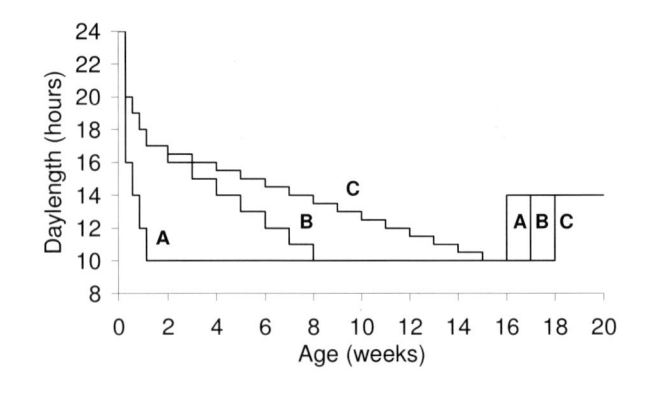

Figure 9.3 *Typical lighting programmes to achieve normal maturity and egg weight (A), to delay maturity by 7 days and increase average egg weight by 1 g (B), or to delay maturity by 14 days and increase average egg weight by 2 g (C).*

increase in egg weight would be produced by giving 14 or 15 weeks of step-down lighting. Typical programmes to achieve these targets are shown in Figure 9.3.

Light intensity

Low light intensity can delay sexual maturity, but it has to be very dim to do so and is unlikely to be a concern in most commercial rearing facilities. Data in Figure 5.2 (page 81) show that sexual development will not be impaired provided the average light intensity at the height of the pullet's head is at least 2 lux.

There are welfare regulations in some countries specifying a minimum light intensity to allow proper inspection of birds and equipment, and it must be right to be able to check birds for signs of ill-health or injury. To do this satisfactorily, the average light intensity needs to be about 5 lux. Provided that pullets can see and be seen, there are no benefits to be gained from operating rearing houses at brighter intensities.

Light leakage

Light-proofing is very important if a producer is to be in complete control of the timing of sexual development. Data in Figure 5.2 (page 81) show that light as dim as 0.05 lux provided 3 hours before and 3 hours after an 8-hour main light period can advance maturity by about a week compared with pullets held on 8 hours. This calls for a very stringent standard of blackout. However, since most commercial pullets will be photostimulated at between 14 and 16 weeks of age and they are not fully responsive to the effects of external increases in daylength until about 9 weeks, the effects of light leakage may not be too devastating in practice.

Light colour

Although experiments have shown that sexual maturity in pullets reared under white or monochromatic red light occurs earlier than in birds reared under pure blue or green light, no significant effects on sexual development have been seen when commercial coloured lamps have been used. There is no benefit to be gained from using other than white light from conventional lamps during the rearing period.

Light source

There is no evidence that one type of lamp is better than any other type for the rearing of growing pullets. Fluorescent lamps have a more expensive capital outlay and generally cannot be dimmed using conventional voltage reduction equipment, but are more cost-effective to operate. In contrast, incandescent lamps are cheap to buy, are easily dimmed using conventional dimmers (should the need arise to control an outbreak of undesirable behaviour), but have higher running costs because less than 10% of their output is emitted as light, the remainder being heat.

Birds are able to see in ultraviolet light (UV-A), and there is evidence from trials with white-feathered turkeys to suggest that the provision of UV-A light, in addition to white light, results in less aggression, better feathering and lower mortality. It is assumed that this is because the birds are able to see the UV-A reflective plumage markings and identify each other as individuals. Further research is being conducted to see if these welfare benefits extend to growing pullets, especially when they are being reared for transfer to natural lighting conditions, where they will be exposed to UV-A radiation. To date, the provision of ultraviolet has entailed separate lighting circuits for the ultraviolet and white light sources, but recently a fluorescent bird-lamp has been introduced that emits both UV-A and white light. However, like most other fluorescent lamps, it cannot be dimmed using simple voltage reduction equipment.

• Conventional lighting during the laying period •

There are several objectives of a lighting programme during the period immediately following the transfer to the laying house. The first is to stimulate feed intake so that the pullets continue to grow at about 70 to 80 g per week. Frequently a transfer can mean a change in feeding and drinking equipment, the introduction of a new diet with different nutrient specifications and texture, changes in light source and intensity, a different temperature and, maybe, a change from a floor system to cages. It is not uncommon during this acclimatisation period for daily feed intake to fall by 10 to 20 g, especially when the birds are not given an increase in daylength. The second aim is to initiate sexual development so that the flock becomes mature at the correct time to produce the combination of egg numbers and weight demanded by the market (a 7-day delay in age at 50% lay means a 5-6 egg reduction in egg numbers but a 1-g increase in mean egg weight). Finally, the lighting programme must encourage an increase in feed intake during the period in which the birds are growing their reproductive organs. If the pullets do not increase their intake at this time they will continue to develop sexually, but do so at the expense of normal growth, and increase the risk of a post-peak dip in egg production.

Once the flock is in lay, daylength will influence the rate of egg production and control the time of egg-laying, but light intensity may affect the welfare of birds, especially through the control of cannibalistic behaviour.

Daylength

On the day that pullets are transferred from the rearing to the laying facilities, they can be without feed or water for many hours and, in the extreme, for more than a day if they are transported over a long distance. It is recommended that they are given a single long day of up to 20 hours to allow them to find feed and water, with access limited to water only for the first 2 or 3 hours. This will have no effect on the timing of sexual development because pullets require exposure to at least a week of long-days to successfully induce sexual maturation in even the earliest maturing members of a flock.

Stimulation of sexual maturation

Once the rearing programme has been defined, the biggest factors controlling the timing of sexual maturity are the size of the increase in daylength and the age at which it is given. Many pullets are now moved at 15 or 16 weeks of age, and, typically, flocks are expected to reach 5% egg production by 18 weeks and 50% before 20 weeks. So, if the first birds are to come into production in the 18th week, and given that the

Table 9.4 *The effect of age at transfer from 8 hours to various new daylengths on age at 50% lay (days) for pullets that would normally reach sexual maturity at 21 weeks of age if maintained on 8-hour daylengths.*

Age at transfer from 8 hours (weeks)	New daylength (hours)				
	9	10	11	12	14
14	139	133	129	126	123
15	140	135	131	129	126
16	141	137	134	132	130
17	143	139	137	135	134
18	144	142	140	139	138
19	146	145	144	143	143
20	147	147	147	146	146

ovary and oviduct take about 14 days to mature, the flock needs to be photostimulated either immediately, or soon after, they have been transferred to the laying house. If they are not stimulated by 16 weeks, the first pullets to mature will do so in response to the rearing base daylength and not to any change in daylength.

The response to an increase in photoperiod is strongest at 9 to 10 weeks of age, but at these ages the pullet has too low a body weight and too small a skeletal frame and, despite successfully becoming sexually mature, such birds will subsequently have extremely poor egg production. It is recommended that growing pullets are not photostimulated before they are 14 weeks of age, nor when their body weight is less than 75% of mature weight.

Table 9.1 gives the effect on maturity of transferring pullets from various base daylengths to different new daylengths at 16 weeks, and shows that the most stimulatory transfer is a move from 10 to 14 hours. Table 9.4 shows that the age of the pullets when they are transferred to the new daylength also has an effect on the age at 50% egg production. It should be noted that photostimulating pullets later than 17 weeks has minimal effect on the timing of 50% lay. It will, however, still encourage an increase in feed intake, influence the time of day at which eggs are laid and may well speed up onset of lay in the later maturing individuals in the flock.

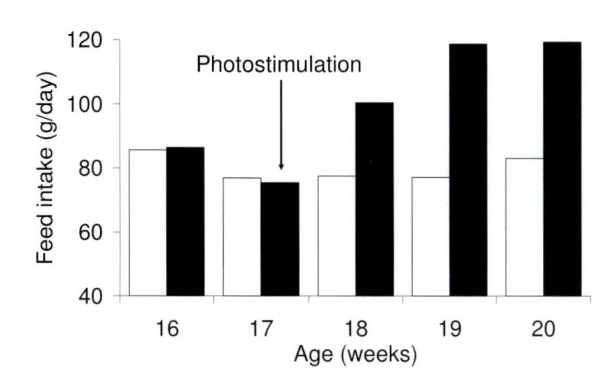

Figure 9.4 *The effect of increasing daylength from 8 to 14 hours at 17 weeks (solid bars) on daily feed intake compared with pullets held on 8-hour days. Data from the University of Bristol.*

Stimulation of feed intake

The other objective of a lighting programme when pullets are transferred to the laying facilities is the stimulation of appetite. Figure 9.4 shows the effect on daily feed intake of transferring pullets from 8 to 14-hour daylengths at 17 weeks of age. These pullets had been moved from the rearing farm to the laying cages at 15 weeks but, even at 20 weeks, the feed intake of the pullets left on 8-hour days was still lower than that at the move. In contrast, feed intake for the pullets transferred to 14-hour days had increased by more than 40 g/day within 2 weeks of photostimulation.

A novel approach to the use of lighting to encourage feed intake during the early part of the laying period is to temporarily provide a 1 or 2-hour light period in the middle of the night. There is no scientific evidence to support this practice, but field experience says it works. This will create an intermittent lighting schedule with two light periods and two dark periods that the hen will use to interpret a subjective day comprising the two light periods and the shorter of the dark periods. To ensure that the subjective day begins with the night-interruption light and ends with the main photoperiod, the dark period before the interruption must be at least one hour longer than the darkness that follows it. This is to ensure that feed consumption, and in particular the intake of calcium, is not limited at the time of day when the hen will have an increased demand for calcium to shell the egg. Typically, a lighting schedule could be 2 hours light, 3 hours dark, 14 hours light and 5 hours dark, which the hen will perceive as a 19-hour day and 5-hour night. The use of such lighting need not be continued beyond about 30 weeks of age, when the night-interruption light is removed. Theoretically, there is a risk that the decrease from a 19-hour subjective day to a conventional 14-hour day will result in some drop in rate of lay, but this appears not to happen in practice. This lighting technique was originally used to solve a shell quality problem in white-egg hybrids during a period of extremely hot weather when feed intake had fallen to a suboptimal level. An alternative in cage operations where the house is lightproof is to change lights to 22.00-12.00 h so that birds are eating in the dark and resting in the heat of the afternoon.

Daylength during the laying period

For many years it was customary to increase the daylength during the laying period to a maximum of 16 or 17 hours. However, the use of such long daylengths stems from the time when poultry houses were not adequately light-proofed, and was to comply with the second golden rule of lighting: 'Never allow the daylength to decrease during lay'. Light-proofing standards have since improved markedly, and there is now no need for daylengths to be any longer than 13 or 14 hours for brown-egg hybrids and 10 or 11 hours for white-egg hybrids to produce maximum egg numbers (see Figure 3.8, page 28).

The laying hen produces more heat during light hours than during darkness and so longer daylengths result in an increase in feed intake to satisfy the extra demand for energy. Daily feed intake goes up by a little over 1 g per bird for each 1-hour of extra daylength, and this matches almost exactly the 1% increase in heat output. As a consequence, average egg weight increases by about 0.25 g for each 1-hour extension of the light period.

In addition to the increased electricity usage, longer periods of illumination have three other negative consequences: thinner shells, more body-checked eggs and higher mortality. The thinning of the shells is more correctly a result of a shorter night than a longer day (see page 28 for physiological explanation) and part of the higher mortality can be attributed to the higher intake of energy that predisposes to an increased number of deaths due to ruptured fatty livers. Body-checked eggs are more prevalent in the morning than later in the day. The condition is thought to be caused by a stress-induced surge of adrenaline causing a strong muscular contraction at a time when the egg shell is still fragile. The shell cracks around the equator and is then repaired with extra calcium carbonate to produce the typical 'equatorial bulge'. Working back from the time that such eggs are laid, it seems that the period associated with the adrenaline surge coincides with lights-out for hens on long daylengths. The end of the light period for hens on shorter days occurs before most eggs have reached the fragile stage and whilst they are still elastic. Although a reduction in daylength has reduced the incidence of these abnormal eggs in experimental conditions, this remedy is not recommended in commercial practice because it will probably result in a drop in production. It is a case of prevention being better than cure: do not put birds on daylengths longer than 14 hours in the first place.

Egg-laying time
It will be seen in Figure 3.12 (page 30) that hens lay their eggs later in the day when they are given longer daylengths and, as a result, a proportion of eggs will be laid before the lights come on when the day is less than 16 hours long. Whereas this is not a major issue for hens housed in cages, it can be a problem in non-cage systems because of the likely increase in the number of eggs laid on the slats or litter when the hens are in darkness. The proportion of 'floor' eggs can be reduced in houses that have automatic nests by having low wattage lights fitted inside the nests and setting them to come on for a 12-hour period starting 2 hours before the main house lights.

Abrupt or step-up increase in daylength
The majority of poultry farmers give, and all breeder management manuals recommend, a relatively modest initial increase of only 1 to 2 hours in daylength at between 14 and 16 weeks of age, followed by a series of 30-minute or 1-hour increases to reach between 14 and 16 hours at 20 to 30 weeks of age. However,

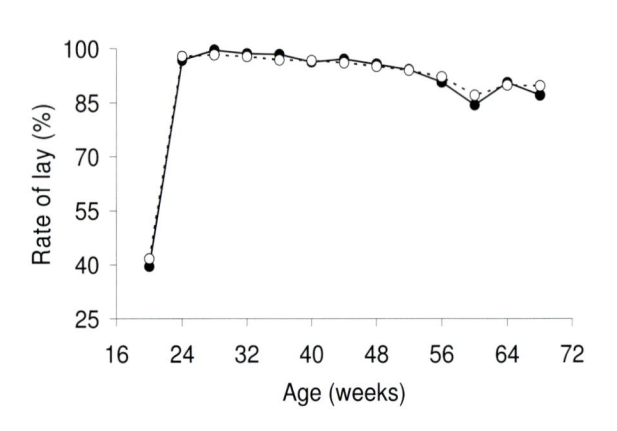

Figure 9.5 *Rate of lay for laying hens transferred abruptly (●) or in a series of 30-minute increments (○) from 8 to 14 hours at 18 weeks of age (unpublished data from the University of Guelph).*

transferring pullets from the rearing to the laying period daylength abruptly has no adverse effects on bird performance; it merely modifies it. The total egg output during the laying period will be the same whether the birds are transferred abruptly to the final daylength or given a step-up lighting schedule. Flocks given a single increment in daylength tend to mature earlier and have a marginally higher peak rate of lay than hens given a series of smaller increases, but the latter birds will generally have better persistency and slightly larger but fewer eggs. Data in Figure 9.5 illustrate the marginal difference in performance between pullets gradually or abruptly increased in daylengths. However, these particular birds were not photostimulated until 18 weeks, and so the difference in age at 50% egg production was small (see Table 9.4, page 126), resulting in only one egg difference in production to 70 weeks of age. Nevertheless, these data do demonstrate that transferring pullets from 8 or 10 hours to 14 hours has no disastrous consequences. The benefits of an abrupt change in daylength include a bigger stimulation of appetite to satisfy the increased demand for nutrients when the pullets are growing their reproductive organs, and fewer eggs being laid before the lights come on in non-caged systems during the early weeks of egg production. Whether increases are given abruptly or gradually will likely be the method that better suits the producer; it will make little difference to the bird.

Welfare regulations

Across the world, an increasing number of poultry industries are being obliged to comply with various pieces of animal welfare regulation that stipulate how they can light poultry. Within the European Union (EU), laying hens must be illuminated so that they comply with the various regulations put into place by member states to enact Directive 88/166/EC, which laid down minimum standards for the protection of laying hens kept in battery cages, and Directive 99/74/EC, which laid down general minimum standards for the protection of laying hens.

- Animals kept in buildings shall not be kept in permanent darkness.
- Where the natural light available in a building is insufficient to meet the physiological and ethological needs of any animals being kept in it then appropriate artificial lighting shall be provided.
- All buildings shall have light levels sufficient to allow hens to see one another and be seen clearly, to investigate their surroundings visually and to show normal levels of activity.
- Where there is natural light, light apertures must be arranged in such a way that light is distributed evenly within the accommodation.
- After the first few days of conditioning, the lighting regime shall be such as to prevent health and behavioural problems. Accordingly it must follow a 24-hour rhythm and include an adequate uninterrupted period of darkness lasting, by way of indication, about one third of the day, so that the hens may rest and to avoid problems such as immunodepression and ocular anomalies.
- A period of twilight of sufficient duration ought to be provided when the light is dimmed so that the hens may settle down without disturbance or injury.

Light intensity

The response of laying hens to light intensity is curvilinear, and so there is no clearly defined intensity above which there is no further increase in egg numbers. However, there appears to be no need to change from the long-standing recommended minimum figure of

Table 9.5 *Effects of light intensity at the feed trough on annual egg production in early and modern hybrids (derived from the curves shown in Figure 5.4).*

Average light intensity (lux)	Early hybrids	Modern hybrids
0.5	245	317
1	249	318
5	258	320
10	262	321
20	266	322
50	272	323

between 5 and 10 lux at the feed trough for optimum performance. Although it has been suggested that selection for egg production over the past five decades has altered the hen's threshold 'requirement' for light intensity, this is probably not the case. A comparison of the effects of reducing light intensity below the recommended threshold for early and modern genotypes reveals that today's hybrids are far more tolerant of low intensities than earlier strains (see Figure 5.4, page 83), but that the threshold has not shifted.

For example, whereas it was predicted 40 years ago that egg production over the laying year would fall by about 13 eggs by reducing the average intensity from 5 to 0.5 lux, egg production from the modern hen is unlikely to drop by more than 3 eggs (Table 9.5).

Surprisingly, brighter light intensities do not stimulate feed intake. In fact, if anything, they marginally suppress it (See Figure 5.6, page 83). Brighter intensities are also associated with a very small, but significant, decrease in egg weight (see Figure 5.5, page 83).

Electricity costs generally increase in proportion to intensity and so, with a marginal increase in egg numbers, a small decrease in egg weight and an increase in feed costs, there are no economic benefits to be gained from providing laying hens with an intensity of more than 5 to 10 lux. Notwithstanding that European Union welfare regulations stipulate these as minimum intensities, it is also common sense and good management to have the light bright enough for the birds and their equipment to be adequately inspected, and so 10 lux appears to be a sensible compromise.

Bird behaviour and mortality

Although brighter light adversely affects mortality, the trend does not continue *ad infinitum* but levels off at about 10 lux. However, the beneficial effects of dimming light on bird behaviour are well established, and EU welfare regulations acknowledge this by permitting a temporary reduction below the stipulated minima to deal with behavioural problems such as pecking and cannibalism.

Twilight

There is evidence to suggest that the provision of a twilight period at the end of the day allows laying hens, especially those in non-caged systems, to settle in a more orderly fashion than when the lights are turned off abruptly. Indeed, there is encouragement to provide such a twilight period in the latest EU welfare legislation. It is also theoretically possible that hens given a period of dim light may produce fewer body-checked eggs if there is less stress involved in settling down for the night. This seems to be a sensible management tool, and can be achieved by installing a separate circuit of low wattage lamps with its time switch set to give 15 minutes of light starting just before the main lights are turned off. For example, 10 W incandescent 'night lights' fitted in every row, but at double the interval of the normal lights (about 6 m), will give an average intensity of about 1 lux at bird level in a floor house (assumes lamps are located about 2 m above the floor). Similar arrangements could also be made in caged laying facilities.

Although natural lighting has both a morning and an evening twilight period, there does not appear to be any need to supply poultry with a gradual dawn.

Light leakage

Most pullets are changed to long daylengths when, or soon after, they are moved into the laying house, and so light leakage is a lesser problem in the laying period than it is in the rearing period.

Light leakage will have an effect on the time of egg-laying, but this will depend on whether it occurs before or after the main light period, and on whether it is brighter or dimmer than 1 lux (see page 84). If the leakage occurs before the house lights come on, egg-laying will take place earlier in the day, even with leakage at very low intensities. However, natural light infiltration after the house lights have gone out will only affect egg-laying if it is brighter than 1 lux at the feed trough in a caged house or at bird height in the perching area of a non-caged house.

Light colour and source

There is no good evidence to indicate that the performance of laying hens is affected in any way by the colour of light or type of lamp used. Poultry see light differently from humans, being more responsive to blue and red. As a consequence, hens perceive light from incandescent lamps to be 10 to 20% brighter than that from fluorescent lamps, even when measurements on a light meter show the intensities to be the same. This is because a light meter is calibrated to the human spectral response curve. Where performance has been claimed to be different from different types of illumination, it has been difficult to separate responses to the light source from those to light intensity (as perceived by the hen). It therefore seems safe to ignore any effects that light source may or may not have on performance and use other characteristics, such as capital outlay, running costs and the ability to dim the lamps, when deciding on the type of lamp to use.

There has been some concern that poultry may see the light from fluorescent lamps as discontinuous. However, flicker is unlikely to be an issue at the light intensities typically provided in laying houses and, in preference tests, laying hens have chosen to spend more time under fluorescent light than under incandescent.

Natural lighting

On free range and in open-sided or windowed housing, hens are exposed to natural lighting. This raises questions about daylength, light intensity and, with the exception of the windowed facilities, exposure to ultraviolet radiation. Additionally, in developed poultry industries, the birds are most likely receiving a combination of natural light, artificial light and darkness.

In most non-controlled environment housing, the problem is the inability to provide the birds with a daylength shorter than the prevailing

Table 9.6 *Schedule of supplemental lighting (GMT times) to maintain a 16-hour daylength throughout the year in naturally illuminated houses at latitude 50°N. The data for latitude 50°S can be found by the addition or subtraction of 6 months; for example, the times for October at 50°N will be the same as those for April at 50°S.*

Month	Natural daylength mid month (h:min)	Artificial lighting (h:min) Morning	Artificial lighting (h:min) Evening
June	03.45-20.15	None required	None required
July	04.00-20.00	None required	None required
August	04.45-19.15	04.00-05.00	19.00-20.00
September	05.30-18.15	03.00-06.00	18.00-19.00
October	06.30-17.15	01.30-07.00	16.30-17.30
November	07.15-16.15	01.00-07.30	16.00-17.00
December	08.00-16.00	01.00-08.30	16.00-17.00
January	08.00-16.30	01.00-08.30	16.00-17.00
February	07.15-17.15	02.00-08.00	17.00-18.00
March	06.15-18.00	02.30-07.00	17.30-18.30
April	05.00-19.00	03.30-05.30	18.30-19.30
May	04.15-19.45	04.00-05.00	19.00-20.00

natural daylength. However, most houses of this type will be in lower latitudes where the longest natural day is unlikely to exceed 14 hours. It has already been shown that pullets respond equally well to an abrupt increase in daylength at photostimulation or to a step-up programme (see Figure 9.5, page 128), and so an abrupt change to a 14-hour daylength when pullets are moved to naturally lit laying houses does not pose a problem. Where birds are exposed to natural lighting throughout the rearing and laying periods, satisfactory egg production can still be achieved by providing a constant 14-hour daylength from day-old to death.

In free range units at higher latitudes, where the longest natural daylength may exceed 16 hours, an immediate transfer to 16 hours at point of lay for pullets reared in lightproof housing on 10-hour days will not adversely affect performance.

The important point about lighting hens exposed to natural lighting is that there must be no decrease in daylength after the longest day. This is easily achieved by supplementing the decreasing natural daylength with artificial lighting, most of which should ideally be given before the beginning of the natural day, with only the final 30 minutes to one hour of supplemental light being given in the evening. With such a programme, the transfer from darkness to artificial light in the morning will be interpreted as dawn, the transfer to the brighter light of the natural light will be perceived as the normal brightening that occurs as the day progresses, and finally the change back to the short period of artificial light will be perceived as dusk, with its lower light intensity creating an evening twilight in which the hens can settle for sleep in an organised way. There is a risk if too much supplemental light is given after the natural day that the reduction in light intensity from natural light, which will amount to tens of thousands of lux, will be such that the hens interpret the transition as dusk and settle down for the night prematurely. Implementing this type of lighting schedule will entail changing the time switch settings at regular intervals as the timing of natural dusk changes. An example of such a programme is given in Table 9.6. The disadvantage of this method of providing supplemental light is that eggs will be laid very early in the morning and this may cause difficulty for poultry keepers who do not have automatic nest boxes or egg-collection systems.

• Intermittent lighting during the laying period •

Intermittent lighting programmes have more than one period of light and darkness in each 24-hour period and can be divided into two distinctly different categories: symmetrical and asymmetrical. Symmetrical programmes (e.g. 2 hours light and 4 hours darkness repeated endlessly) do not allow the bird to identify a part of the 24-hour cycle as 'night' and the remainder as 'day', whereas asymmetrical programmes (e.g. 2 hours light, 4 hours darkness, 10 hours light and 8 hours darkness) do allow this distinction to be made.

Asymmetrical programmes

These programmes have two or more periods of light and darkness but, as the name suggests, the periods of darkness are not of equal sizes. The hen identifies the longest period of darkness as 'night', and treats the remaining part of the cycle, be the lights on or off, as a subjective day. The hen responds to such a programme as if it were a conventional night and day (see pages 57-58). The programmes were originally developed in America to save electricity without losing performance, and two examples are the system developed at Cornell University and *Biomittent®* lighting, a regime designed by the Ralston Purina Company. The Cornell programme provides an 8-hour photoperiod preceded by a secondary 2-hour photoperiod starting 6 hours before the main one. The programme is more simply described as 2L:4D:8L:10D, where L = hours of light and D = hours of darkness. The hen responds as if it was a conventional 14L:10D schedule and so, provided feed intake is satisfactory, it can be safely used from the beginning of the laying period. The *Biomittent* system aimed to save even more electricity than the Cornell programme by segmenting each hour of a 16-hour 'day' into 15 minutes light and 45 minutes darkness, though the final hour is made up of 15 minutes light, 30 minutes darkness and 15 minutes light to make a whole number of hours This can be described as 15(15minL:45minD):15minL: 30minD:15minL):8D. Even though, like the Cornell, it mimics a conventional day and night, it is often not introduced until after the hens have reached peak egg production because the savings in feed can be such that protein intake is insufficient to support maximum egg output. Whilst it can be used from 18 weeks of age by fortifying the diet to ensure an optimum intake of amino acids, it is most important at any age that it is introduced gradually. The day-time hours should comprise 45 minutes light and 15 minutes darkness for the initial 1 or 2 weeks, then a period of 30 minutes light and 30 minutes darkness before changing to 15 minutes light and 45 minutes darkness when the feed intake is satisfactory. Data in Figure 4.3 (page 59) show that egg production under asymmetrical lighting is similar to that expected for the equivalent conventional lighting programme provided there is no limitation of protein intake.

Asymmetrical performance

Typically, under most asymmetrical intermittent programmes, there is a reduction in feed intake, an improvement in feed conversion efficiency, a smaller end-of-lay body weight and lower mortality (Table 9.7). The savings in feed intake are generally greater for *Biomittent* type programmes than for the Cornell because of the smaller total period of illumination. Shell quality for hens given asymmetrical lighting is at least as good as that for conventionally illuminated hens, and eggs are laid at the same time as if the birds had been given a conventional day and night programme.

Asymmetrical programmes fully comply with current EU welfare regulations, provided the longest dark period is at least 8 hours, because they allow the hen to recognise a 24-hour cycle.

Symmetrical or short-cycle programmes

This type of intermittent lighting is made up of light-dark cycles that repeat at less than 24 hours. The programmes, which were originally developed in France to improve shell quality and egg weight, normally repeat every 6 or 8 hours. The most commonly used system is a 6-hour one with 3-hour periods of light alternating with 3 hours of darkness. However, with these programmes, the hen is unable to form a day and night, even though the cycle fits exactly into a 24-hour period, because all of the dark periods are the same size. The bird then

Table 9.7 *Egg laying performance to 72 weeks of age for hens given Cornell or Biomittent asymmetrical lighting compared with performance for conventionally illuminated controls. The data are from Universities of Reading and Bristol and North of Scotland College.*

Trait	Lighting programme Cornell	Lighting programme Control	Lighting programme *Biomittent*	Lighting programme Control
Egg production (%)	79.2	78.6	78.9	79.0
Egg weight (g)	62.5	62.9	63.3	63.4
Egg output (g/d)	49.8	49.9	50.7	50.8
Feed intake (g/d)	119.0	122.4	114.2	118.7
Feed conversion (g/g)	2.40	2.48	2.26	2.34
Shell thickness (specific gravity)	1.080	1.080		
Shell thickness (mg/cm)			74.2	70.8
End body weight (kg)	2.26	2.36	2.17	2.17
Mortality (%)	2.3	3.8	3.5	4.5

'free runs' and responds to a natural rhythm that controls many physiological functions, including ovulation, and is longer than 24 hours. This underlying rhythm is over-ridden when the bird is able to form a night and day, and so is not seen in conventionally lit birds or those given asymmetrical intermittent lighting.

Although pullets may be transferred to symmetrical programmes soon after they are moved into the laying house, the change is not normally made until after peak egg production. This is to prevent eggs being laid randomly throughout the 24 hours. The cue for initiating the ovulatory cycle is the transition from light to darkness at the end of the day (dusk), and, whereas conventionally lit birds have only one such cue, hens given 6-hour short-cycle programmes will have a choice of four, and individuals will select their own light/dark interface at random. Providing an initial period of conventional lighting and then having one of the short-cycle photoperiods ending at the same time as the lights-out of the conventional programme will help synchronise the flock initially to a common 'dusk'.

Many trials and extensive commercial usage have shown that this type of intermittent programme results in an improvement in shell quality, an increase in egg weight, but a corresponding reduction in egg numbers when introduced early in the laying year (Table 9.8). The better shell quality and increased egg weight under symmetrical lighting are achieved by releasing the hen from the confines of the 24-hour solar day, so lengthening its ovulatory cycle. Yolk formation occurs at a constant rate and so, as a consequence of the longer ovulatory cycle, the egg contents will be bigger and the shell thicker (because of the longer time spent in the shell gland). Egg mass output is similar to that for conventional lighting. Feed intake is reduced and feed conversion improved compared with conventional lighting. Feed consumption can be further reduced by using cycles of 2.5L:3.5D, 2L:4D or 1.5L:4.5D. Mortality is consistently lower with symmetrical programmes than with conventional systems.

A special form of symmetrical lighting was developed at the University of Reading to combine the feed saving and improved feed conversion benefits of *Biomittent*, the thicker shells and bigger egg weight of a symmetrical intermittent lighting programme, and the lower mortality of both. The Reading system uses a 1-hour cycle with 15-minute light periods alternating indefinitely with 45-minute dark periods. Data in Table 9.9 show that the aims were achieved, but that they were at the expense of egg numbers, with hens on the novel programme laying 7 fewer eggs to 72 weeks than the conventionally lit controls that had been stepped up to a 15-hour daylength by 27 weeks of age. Shell quality was not measured in this comparison, but, in an earlier trial, shell thickness for the Reading system was similar to that for a 6-hour symmetrical 3L:3D

Table 9.8 *Egg laying performance to 72 weeks of age for hens given short-cycle intermittent lighting compared with performance for conventionally illuminated controls. The data are pooled from research centres in UK and France.*

Trait	Short-cycle	Normal lighting
Egg production (%)	78.0	81.0
Egg weight (g)	64.0	62.5
Egg output (g/d)	49.3	50.2
Feed intake (g/d)	115.2	121.8
Feed conversion (g/g)	2.35	2.44
Mortality (%)	2.3	3.8

Table 9.9 *Egg laying performance to 72 weeks of age for hens given the Reading system (15minL:45minD repeated endlessly) compared with performance for conventionally illuminated controls.*

Trait	Reading system	Step-up to 15 hours
Egg production (%)	78.2	80.9
Egg weight (g)	65.1	63.8
Egg output (g/d)	50.3	50.5
Feed intake (g/d)	116.3	123.4
Feed conversion (g/g)	2.28	2.39
Mortality (%)	3.6	4.9

programme, and both types of lighting produced significantly thicker shells than a *Biomittent* programme. Fifteen minutes is insufficient time to collect eggs and inspect hens properly, but the time switch can be temporarily overridden to give a long enough period of light for servicing the birds.

Red mite control

All short-cycle repeating programmes reduce red mite infestations. It is thought that this may be a consequence of the dark-periods being so short that the mite are either prevented from sucking blood or cannot ingest enough to reproduce. However, symmetrical programmes cannot be used in EU countries because they contravene welfare regulations.

Which intermittent programme to use?

All intermittent programmes have some advantage over conventional lighting, but how does one decide which type to use? It depends on several factors, including how well the house is light-proofed, whether the unit is manual or automated, prevailing welfare regulations, and what aspect of performance is economically most important. The following lists of characteristics of the Cornell, *Biomittent*, French symmetrical and Reading systems give the benefits of each type relative to conventional lighting.

Cornell (2L:4D:8L:10D)
- Similar egg numbers
- Similar egg weight
- Similar egg output
- Similar shell quality
- Lower feed intake
- More efficient feed conversion
- Lower mortality
- Reduced electricity usage
- Good light-proofing not essential
- Permitted under EU welfare regulations

***Biomittent* (16(15minL:45minD):8D)**
- Similar egg numbers
- Similar egg weight
- Similar egg output
- Similar shell quality
- Larger feed saving than Cornell
- More efficient feed conversion
- Lower mortality
- Greater electricity saving than Cornell
- Good light-proofing not essential
- Permitted under EU welfare regulations

French symmetrical (3L:3D repeating)
- Fewer egg numbers
- Heavier egg weight
- Similar egg output
- Improved shell quality
- Lower feed intake
- More efficient feed conversion
- Lower mortality
- Control of red mite infestation

- Reduced electricity usage
- Good light-proofing essential
- Servicing may be difficult
- Not allowed under EU welfare regulations

Reading symmetrical (15minL:45minD repeating)
- Fewer egg numbers
- Heavier egg weight
- Similar egg output
- Improved shell quality
- Lower feed intake than French system
- More efficient feed conversion
- Lower mortality
- Control of red mite infestation
- Greater electricity saving than French system
- Good light-proofing essential
- Servicing requires overriding the time switch
- Not allowed under EU welfare regulations

Flexibility of intermittent programmes

All forms of intermittent lighting programme are completely reversible, so if conditions change, such as a reduced demand for large eggs, a flock can be safely returned to conventional lighting without disruption to the laying pattern.

Insect control

On the previous page we reported that short-cycle intermittent programmes are able to reduce infestations of red mite, but asymmetrical programmes have also been shown to control insect populations. Notwithstanding that it was demonstrated in closed turkey housing, the use of a programme containing three 1-hour periods of light at the start, middle and end of a 10, 12 or 14-hour day successfully reduced the house-fly population (Table 9.10). There is no reason not to assume that such programmes would control fly populations in any type of lightproof poultry housing.

Table 9.10 *Mean fly numbers in turkey houses under 10, 12 or 14-hour conventional and asymmetrical intermittent lighting*

Location	Lighting system	
	Conventional	Intermittent
Feeders	16	4
Jug traps	9	6
Tapes	162	14
Spot cords	68	10
Spot papers	33	6
Total	**289**	**40**

• Ahemeral lighting during the laying period •

Ahemeral, or non-24 hour, lighting regimes were in vogue in the 1960s, but are rarely, if ever, used today. They were used to increase egg weight and improve shell quality, and achieved this by extending the interval between eggs. The surge in luteinizing hormone that triggers the release of the yolk from the ovary (ovulation) can only occur once in each light-dark cycle, and so the minimum interval between ovulations, when hens are given 27-hour rather than 24-hour daylengths, will be 27 hours. This results in the next yolk being larger, and the egg spending longer in the shell gland to produce a larger, thicker shell. Although these benefits are still required from time to time, they can more easily be produced by transferring birds to one of the short-cycle symmetrical lighting programmes such as 3 hours of light alternating with 3 hours of darkness.

A disadvantage of ahemeral programmes, such as a 27-hour daylength, is that by the middle of each 9-day cycle the hens' day is our night, and *vice versa*, thus making servicing of the birds very difficult. However, this can easily be overcome by using a bright-dim, instead of a light-dark, cycle that has a 10:1 ratio of the bright and dim intensities. This involves the use of a light intensity during the bright period that is 10 times brighter than would normally be used in a poultry house if the dim intensity is bright enough for servicing the birds. For example, if 5 lux intensity is necessary for servicing, the brighter light period would need to be 50 lux, and this may have an adverse effect on bird behaviour. Also, the extra electricity required for this brighter intensity and the continuous use of lights will increase electricity and lamp replacement costs, thus offsetting some of the financial benefit of larger eggs and better shell quality. A further downside of ahemeral lighting is that when it is introduced early in the laying period to increase egg weight there will be a loss in egg numbers. If used towards the end of the laying year, when shell quality is a problem, ahemeral lighting may actually increase rate of lay. A final problem for poultry producers in the EU is that ahemeral lighting is now prohibited because the hens are not allowed to follow a 24-hour rhythm and, in the case of bright:dim programmes, the hens do not have a period of darkness. The use of ahemeral lighting is therefore now part of the history of lighting laying hens, especially when it requires lightproof housing which is not always available in the parts of the world that do not yet have to comply with welfare regulations.

There is a theoretical future potential for ahemeral cycles of less than 24 hours. Maximum rate of lay is achieved by providing a cycle that matches a hen's natural interval between eggs (see Table 4.6, page 66). One of the consequences of selection for increased egg production is that hens now have shorter intervals between eggs because they can form an egg more quickly than earlier hybrids. It would be possible to further increase egg production if hens could be given a day that was 23 or 23.5 hours long. Clearly, this is fairly impractical, probably not cost-effective and is illegal with the EU, but it is biologically interesting.

Further information on the mechanisms involved in ahemeral lighting is given in chapter 4, pages 65-69.

10 LIGHTING FOR BROILER BREEDERS

Lighting for broiler breeders is complicated because they cannot simply be treated as big chickens. Unlike laying hens, broiler breeders are still seasonal breeders and so, like turkeys, must be given a period of short days before they can be photostimulated into maturity. The response to lighting programmes also varies with the body weight of the bird and so different weight birds must be stimulated at different ages. This chapter describes typical recommended lighting programmes and discusses the consequences of recent research findings for the future lighting of broiler breeders.

• Lighting during the rearing period •

As with growing pullets, it is standard practice to give broiler breeders 2 days of 23 hours of light or continuous illumination for them to settle in and find the feeding and drinking equipment. Thereafter, lighting programmes for broiler breeders are quite different. Much of the control of sexual maturity comes from the restricted feeding programme, and so the main function of the lighting programme is to make the bird responsive to an increase in daylength, not to delay maturity. A broiler breeder is still a seasonal breeder, which means that it is hatched photorefractory and unable to respond to a light increase until this state has been dissipated. Photorefractoriness is a condition that normally prevents a wild bird from breeding in the year in which it hatches (pages 16-17). Most wild birds hatch in the spring but do not become sexually mature in the late summer, even though they are of a mature body size and the daylength is similar to that when their parents bred. This mechanism prevents animals from trying to rear young when environmental conditions are not favourable for doing so. True seasonal breeders have photorefracroriness dissipated by the short days of winter. Although broiler breeders do not express this extreme form of photo-refractoriness, they still need a period of 8-hour daylengths to become fully photosensitive. In nature, photorefractoriness is dissipated by about 2 months of short days, but the level of feed restriction commonly applied to broiler breeders delays the dissipation of photo-refractoriness so that it takes 4-5 months of 8-hour days for all birds in a broiler breeder flock to become responsive to a light increase. In contrast, *ad libitum* fed growing pullets are fully responsive by about 9 weeks of age.

Daylength for lightproof housing
When broiler breeders are reared on any daylength between 6 and 10 hours and then transferred to longer days at 20 weeks there will be only 3 or 4 days difference in age at 50% lay (Table 3.2, page 35), no more than 3 eggs difference to 60 weeks and less than 0.5 g difference in egg weight. Typically, broiler breeder companies advocate a series of decreases in daylength from the initial long days to 8 hours by 2 weeks of age (Table 10.1). These almost daily changes of the time switch have negligible effect on sexual maturity or subsequent egg production, and are completely unnecessary from a lighting point of view. However, it may be felt from a stockmanship standpoint that a gradual decrease to 8 hours has management benefits over an abrupt decrease. If these benefits are not important, the daylength may be safely reduced by moving the time of lights-on to 8 hours in one or two steps to reach 8 hours by 4 or 5 days of age.

Daylength for non-lightproof housing
It is common practice when pullets are exposed to variable natural daylengths in houses that are open sided or poorly light-proofed to give some form of step-down lighting programme to reach a daylength that is equal to the natural days expected at 19 or 20 weeks. However, when such programmes are used there will still be

Table 10.1 *Typical lighting programmes recommended by primary breeding companies for broiler breeders maintained in lightproof facilities.*

Age (days)	Cobb 500	Ross 308	Arbor Acre	Hubbard Classic
1	24	24	23	22
2	20	19	19	20
3	20	16	16	18
4	18	14	14	17
5	18	12	12	16
6	16	11	11	15
7	14	10	10	14
8	13	9	9	13
9	12	8	8	12
10	11	8	8	11
11	10	8	8	10
12	9	8	8	9
13-133	8	8	8	8
134-140	12	8	8	8
141-147	13	11	8	10
148-154	14	12	12	11
155-161	15	12	12	12
162-168	16	13	13	13
169-175	16	13	13	14
176-182	16	14	14	15
183-189	16	15	14	15.5
from 190 onwards	16	15	15	16

Information in this table has been obtained from management guides issued by the following primary breeding companies (current in 2005).

Cobb-Vantress	Aviagen	Hubbard
Siloam Springs	Edinburgh	Duluth
U.S.A.	Scotland	U.S.A.

some seasonal variation in age at 50% egg production (Figure 3.27, page 37).

Lighting for naturally increasing daylengths

Research at the University of KwaZulu-Natal showed that there was no significant effect on age at 50% egg production for broiler breeders given simulated naturally increasing or decreasing daylengths (between 10 and 14 hours) and those maintained from a young age on 14-hour days. However, the birds given constant 14-hour days subsequently laid between 6 and 10 fewer eggs to 60 weeks than either of the 'naturally illuminated' groups (Table 10.2). The similar ages at sexual maturity for all three lighting treatments demonstrate the dominant role that the feeding programme plays in the control of sexual development in broiler breeders.

This trial showed the slightly inferior egg production of broiler breeders reared on long days, compared with birds given naturally decreasing or increasing daylengths, but other work from the University of KwaZulu-Natal shows an even greater disparity when the performance of birds reared on long days is compared with that of birds reared on 8-hour days and transferred to 16-hour days at 20 weeks. Table 10.3 illustrates the retarding effects of the 16-hour days during rearing, the adverse effects on subsequent egg production and the undesirable increase in egg weight.

These research findings show the importance of giving broiler breeders an adequate period of short days during the rearing period to hasten the dissipation of photorefractoriness, and question the correctness of rearing birds on long days when they would otherwise be exposed to naturally increasing daylengths. Although the use of long days to prevent pullets having an increase in daylength during the rearing period may be sound advice for *ad libitum* fed egg-type hybrids, it does not appear to be appropriate for broiler breeders. This is because sexual maturity is markedly delayed when broiler breeders are maintained on daylengths that are longer than 11 hours during the rearing period.

Table 10.3 *Age at 50% lay, and total egg numbers and average egg weight to 60 weeks for broiler breeders reared on 8-hour days and transferred to 16-hour days at 20 weeks or maintained on 16 hours throughout.*

Trait	Rearing daylength	
	8 hours	16 hours
Age at 50% egg production (days)	193	211
Eggs per bird	169	154
Egg weight (g)	67.5	68.2
Egg output (kg)	11.4	10.5

Table 10.2 *Age at 50% egg production and laying performance to 60 weeks for broiler breeders grown to breeder body weight targets and given simulated increasing or decreasing natural daylengths, or maintained on 14-hour days to 20 weeks and then transferred to 16 hours.*

Trait	Increasing from 10 hours	Increasing from 11 hours	Decreasing from 14 hours	Constant 14 hours
Age at 50% egg production (days)	209	210	209	211
Eggs per bird	150	146	150	140
Average egg weight (g)	70.0	69.8	69.4	68.9
Egg output (kg)	10.49	10.21	10.42	9.68
Feed to produce each egg (g)	337	341	333	357

Lighting to 20 weeks

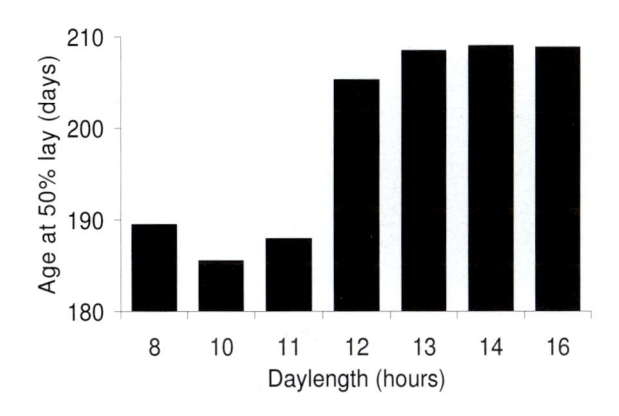

Figure 10.1 *Effect of constant daylength throughout the rearing and laying periods on age at 50% egg production in broiler breeders grown to breeder body weight targets. (data from the University of KwaZulu-Natal).*

The sexual maturity data in Figure 10.1 clearly show that rearing daylengths can be divided into two distinct classes; those that are interpreted as short days (11 hours or less) and allow the birds to quickly become photoresponsive, and those that are stimulatory (12 hours or longer) and retard sexual development.

Variable or naturally decreasing daylengths
When broiler breeders are exposed to variable or naturally decreasing daylengths during the rearing period, simply use the natural pattern and increase to at least 14 hours at 20 weeks.

Light intensity
There are no reports in the scientific literature for the effects of light intensity during the rearing period on the performance of broiler breeders. Nevertheless, there is little variation in the advice given by the main primary broiler breeding companies and it also agrees with the scientifically backed recommendations for rearing egg-type hybrids. Following a period of bright light for the first 2 or 3 days to allow the birds to find feed and water, it is suggested that the light intensity should be reduced to between 5 and 10 lux at bird-head height for the remainder of the rearing phase.

Light colour and source
In recent years, there has been a lot of interest in the use of coloured lights for rearing broiler breeders. However, there are no research findings to back the claims for improved performance from the use of such lamps and producers are advised to think twice before replacing their current lamps with coloured compact fluorescent. It seems that the research findings that showed red light advances sexual maturity have been misinterpreted to indicate advantages for the use of coloured lamps in domestic fowl. The original research with Mallard drakes showed that truly monochromatic red light was more sexually stimulatory than other colours of light (pages 11 and 12). However, the light from coloured compact fluorescent lamps is far from monochromatic. Data in Table 10.4 and Figure 10.2 show that such lamps emit a broad spectrum of light A typical 'red' fluorescent lamp produces less than 20% of its output in the red band and has a spectral range that is not too dissimilar to that of a warm-white fluorescent (Figure 10.2).

Chapter 10 Lighting for broiler breeders

Table 10.4 *Typical spectral emissions (% in each band) from coloured compact fluorescent lamps. Nominal colour in bold type.*

Colour/wavelength band	Blue	Green	Red
UV-A <380 nm	3.0	0.0	0.0
Violet 380-435 nm	19.2	6.1	6.9
Blue 435-500 nm	**64.7**	18.5	5.4
Green 500-565 nm	10.4	**49.8**	6.0
Yellow 565-600 nm	1.6	14.2	11.9
Orange 600-630 nm	0.4	7.5	51.6
Red 630-780 nm	0.8	3.9	**18.2**

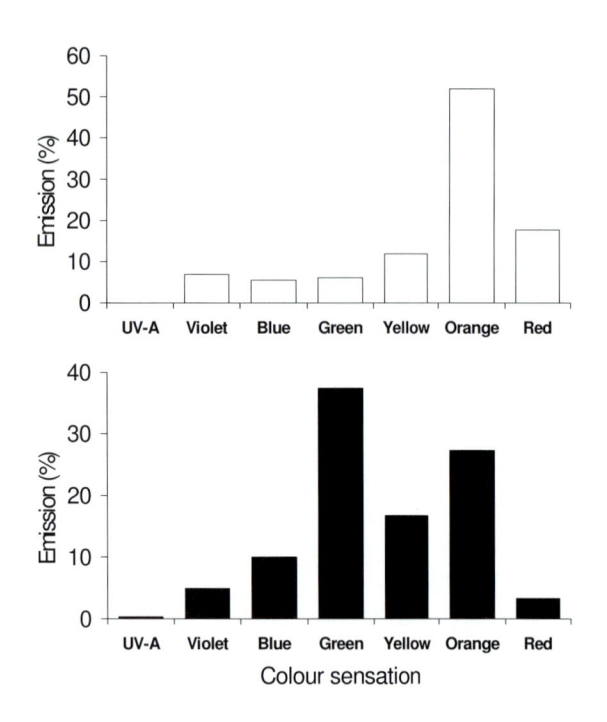

Figure 10.2 *Spectral emissions from typical compact red (open bars at top) and warm-white (solid bars at bottom) fluorescent lamps.*

There is no evidence to suggest that one lamp type is better than any other for rearing broiler breeders. The choice will therefore depend on availability, capital outlay, running costs and the ability to dim using conventional voltage reduction equipment.

Males

Unpublished data from the University of KwaZulu-Natal suggests that the response of male broiler breeders to daylength during the rearing period is similar to that of females, so there is no reason why the two sexes cannot be given the same lighting programme.

• Lighting during the laying period •

Daylength for lightproof housing
Age at photostimulation

The lighting programmes in Table 10.1 show that broiler breeders are typically maintained on 8 hours until 19 to 21 weeks of age and then given a transfer to between 10 and 12 hours followed by a series of increments to reach 15 or 16 hours at between 23 and 27 weeks. However, estimates for age at 50% egg production, using typical body weights and the prediction equation on page 36, show that there is unlikely to be more than a week's difference in maturity among the programmes (Table 10.5). The relatively small differences in the response to each of the four lighting programmes is a result of the size of the light increase and the age at which it is given interacting with the body weight at photostimulation (the effects are quantified on page 36).

When pullets are transferred to a long daylength at an older age, the age-related delay in sexual maturity will be offset by the stimulatory effect of photostimulating at a heavier body weight. Additionally, the effects of these increases will vary with the rate at which the daily feed allocation is increased after the birds have been transferred to long days.

Research at the University of KwaZulu-Natal indicates that broiler breeders that have been grown according to body weight targets recommended by the breeding company should not be photostimulated until they are at least 19 weeks of age. This is because, at younger ages, some birds will still not be responsive to a light increase, even if their body weight is on target and the flock is uniform. If broiler

Table 10.5 *Predicted age at 50% egg production (days) for broiler breeders transferred from 8 to 10, 11 or 12 hours light at 19, 20 or 21 weeks of age and weighing 2.16 kg at 20 weeks using the equation on page 36.*

Age at transfer from 8 hours	Daylength (hours)		
	10	11	12
19 weeks	190	187	184
20 weeks	190	187	184
21 weeks	193	189	187

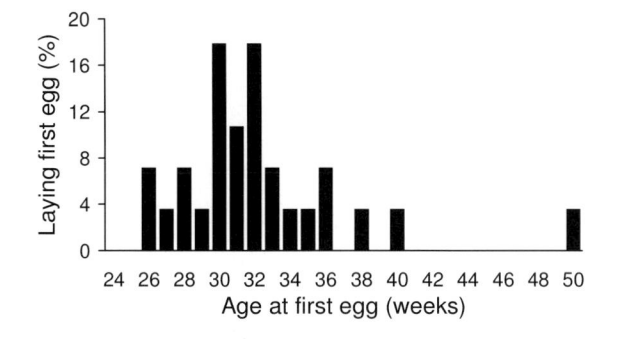

Figure 10.3 *Proportion of broiler breeders laying their first egg after a transfer from 8 to 16 hours at 14 weeks of age (data from the University of KwaZulu-Natal).*

breeders are transferred to long days when they are younger than 19 weeks, sexual development in a proportion of birds may be delayed by weeks or even months. Figure 10.3 shows the proportion of pullets that laid their first egg during weeks 26 to 50 after they had been transferred from 8 to 16 hours at 14 weeks.

Underweight or uneven flocks

The first increase in daylength should be delayed by a week if the mean body weight is below breeder specifications or if the flock is uneven (more than 5% of the flock outside the range ± 20% of the average). This is because maturity is retarded and subsequent egg production depressed if a pullet is photostimulated when its body weight is less than 2.0 kg.

Initial increase and maximum daylength

The initial increase in daylength from 8 hours that is recommended by the primary breeding companies varies from 2 to 4 hours, and is followed by various increases to a maximum of 15 or 16 hours. The effect of the initial increase in daylength for broiler breeders transferred from 8 hours at 20 weeks is very similar to that for commercial egg layers (Figure 3.22, page 35, for comparison). The data in Figure 10.4 show that the largest advances in age at 50% egg production are achieved by a transfer to 14 or 16 hours. Because of these differences in the onset of lay, birds that have been transferred to the longer daylengths produce more eggs to 39 weeks (Figure 3.25, page 37), but other research has consistently shown that broiler breeders given long days have inferior egg production after peak rate of lay, resulting in fewer eggs being laid to 60 weeks of age, despite the earlier maturity. For example, the 16-hour flock in Figure 10.5 produced 5 fewer eggs per bird to 60 weeks, despite maturing 5 days earlier than the birds held on 11 hours light per day.

The inferior egg production is caused, in part, by longer, more stimulatory daylengths advancing the age at which individual birds become photorefractory and go out of lay, and this depresses the average rate of lay for the flock. Incidentally, there is no compensatory increase in egg weight associated with this reduction in egg numbers. Another factor that may contribute to the poorer egg production is the higher energy demand for maintenance when birds are in light rather than in darkness.

Data for laying hens shows that birds on 16-hour days expend 5% more energy than those on 11 hours, and so, because broiler breeders are given fixed quantities of feed, less

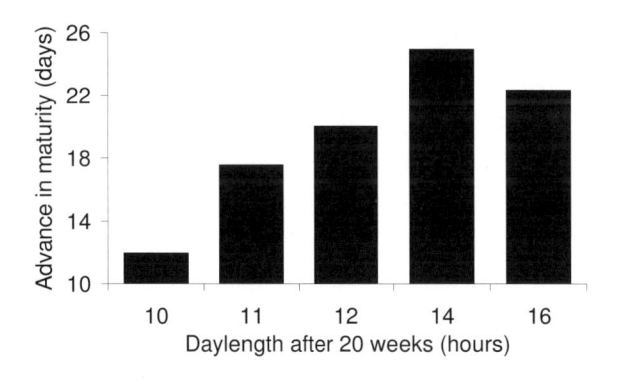

Figure 10.4 *Advance in age at 50% egg production, compared with holding on 8 hours, achieved by transferring broiler breeders from 8 hours to various daylengths between 10 and 16 hours at 20 weeks (data from University of KwaZulu-Natal).*

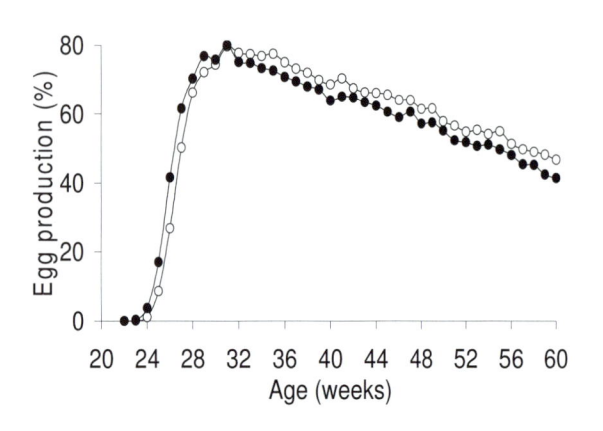

Figure 10.5 *Rate of egg production for broiler breeders transferred from 8 hours to 11 (○) or 16 (●) hours at 20 weeks (data from the University of KwaZulu-Natal).*

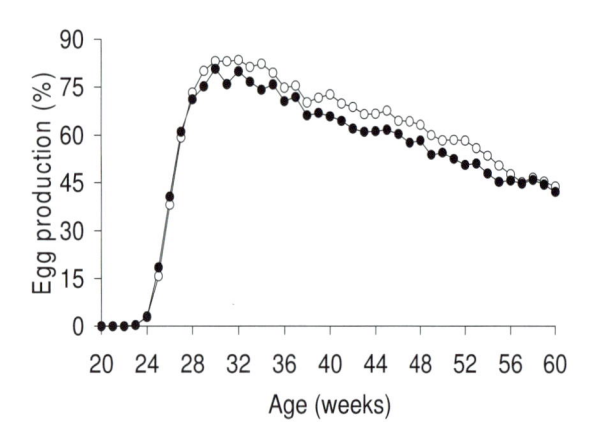

Figure 10.6 *Rate of egg production for broiler breeders transferred from 8 hours to 12 (○) or 16 (●) hours at 20 weeks (data from University of KwaZulu-Natal).*

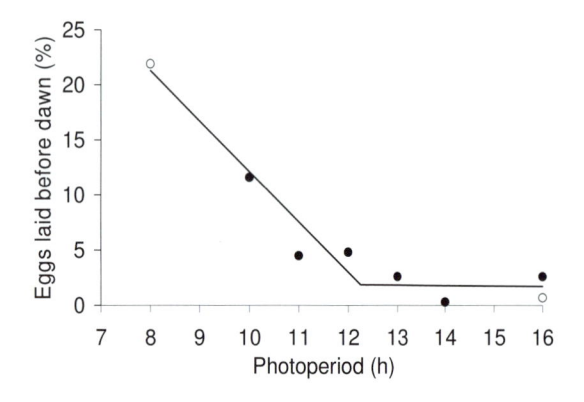

Figure 10.7 *Percentage of eggs laid before lights-on by broiler breeders given daylengths of between 8 and 16 hours (data from the University of KwaZulu-Natal).*

energy is available for egg production. Data in Figures 10.5 and 10.6 for birds transferred from 8-hour days and in Figure 3.18 (page 33) for birds given constant daylengths from a young age show that extending the daylength to 16 hours depresses rate of lay post-peak, resulting in 5 to 10 fewer eggs to 60 weeks. It is interesting to note that the rate of lay after 55 weeks for birds transferred to 12 hours was similar to that for the hens on 16-hour days (Figure 10.6). This is likely to have been a combination of the 12-hour birds tailing off and the 16-hour birds improving due, possibly, to some birds coming back into lay after having had a pause in the middle of the laying period; this also occurs in turkeys (page 18). The difference between the rates of lay for the 11 and 12-hour birds in the final weeks of the laying cycle probably reflects the more stimulatory nature of 12-hour days.

Although this suggests that increases beyond 11 or 12 hours are unwarranted, there is a downside to using these shorter daylengths, and that is the likelihood that more eggs will be laid on the floor because hens on shorter daylengths begin laying before the start of the light period (page 34). For example, it can be expected that hens on 11-hour days will lay about 5% more eggs before dawn than hens on 16-hour daylengths (Figure 10.7). However, the risk of 'floor-laying' can be reduced in houses fitted with automatic nests by installing low wattage (less than 10 W) incandescent lamps inside the nest boxes and turning them on for a 12-hour period starting 2 hours before the main house lights come on.

Step-up or abrupt increase?

Birds can cope very well with an abrupt increase in daylength, and it is not essential to give a step-up programme to achieve optimum performance. We have seen earlier that there is no need to give daylengths that are longer than 11 or 12 hours to maximise egg production and so, if this policy is implemented, there will be no further increases after the initial increment and such flocks will have received an abrupt increase in daylength. If broiler breeding operations continue to provide a maximum daylength of 15 or 16 hours, as recommended by many breeding companies, the difference in performance between going to 16 hours abruptly and reaching it over several weeks is in the pattern of egg production. Whereas abruptly photostimulated flocks mature a little earlier (because 8 to 16 hours is more stimulatory than 8 to 11 or 8 to 12 hours, Figure 10.4) and have a slightly higher peak than those given a series of increments, birds given a step-up programme have marginally better persistency. However, the total egg output is likely to be very similar for the two flocks. Figure 10.8 shows the rates of lay for broiler breeders reared on 8 hours and either changed abruptly to 16 hours at 19 weeks or to 12 hours followed by weekly increments of 1 hour to reach 16 hours at 23 weeks.

Further increases in daylength
It has been suggested that it could be beneficial not to give supplementary increases in daylength immediately after the initial increment but to hold them back until later in the laying cycle. The thinking is that once the pullets have been stimulated into sexual maturity by an increase to 11 or 12 hours, there is little point in 'wasting' the effects of further increases. Indeed, if the initial daylength is long enough to initiate sexual development, and further increases have no effect on feed intake because this is fixed by the feeding programme, the extra hours of light can have no effect other than to influence the time of day when the eggs are laid. It is thought that the broiler breeders' sensitivity to light may change with age, and that they would be able to respond, in terms of increased rates of lay, to a series of increments in daylength given after about 40 weeks of age. Unfortunately, this does not appear to be the case, and transferring broiler breeders from a moderately stimulatory daylength, such as 11 hours, to a 16-hour day in 15-minute weekly or 1-hour monthly increments has a depressive and not stimulatory effect on egg production (Figure 3.26, page 37). Since broiler breeders are 'seasonal breeders', changing to a more stimulatory daylength seems to simply accelerate the end of the 'breeding season'.

Daylength for non-lightproof housing

A large proportion of the world's broiler breeders are housed in facilities that are either open-sided or insufficiently light-proofed to effectively operate a controlled light programme. Although it is highly desirable that broiler breeders are kept in lightproof facilities during the rearing period, so that they can be given short days to dissipate photorefractoriness as quickly as possible, moving them from light-proof into non-lightproof houses for the laying period does not create too many problems provided they are not moved before 19 or 20 weeks of age. If moved at a younger age and the natural daylength is more than 10 hours then some of the birds will not respond positively to the increase in daylength but will have their maturity markedly delayed (see Figure 10.3, page 141) and the flock will be uneven. When there is no alternative but to move birds earlier than 19 weeks of age, it is advisable to rear them on a 10-hour day to reduce the increment when they are transferred to the natural lighting conditions.

When both the rearing and the laying house are poorly light-proofed, and the birds have been reared on a naturally decreasing daylength, simply increase to at least 14 hours at 20 weeks. If the natural daylengths have been increasing, just follow the natural lighting pattern to mid summer, but ensure that the daylength does not decrease once the birds have started egg production.

Light intensity

There are no scientific publications describing the responses of broiler breeders to light intensity. However, recommendations in the primary breeding companies' management

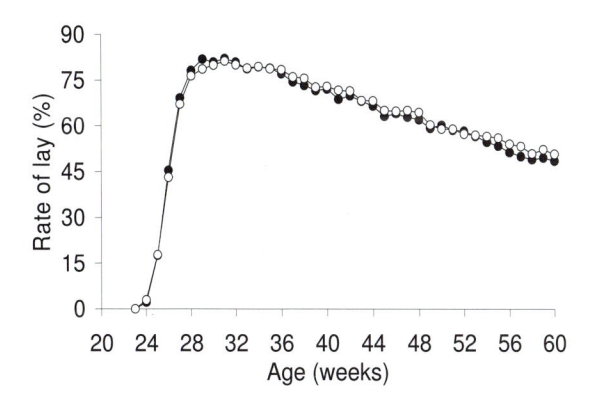

Figure 10.8 *Egg production for broiler breeders transferred abruptly from 8 to 16 hours (●) or from 8 to 12 hours (○) at 19 weeks followed by weekly increments of 1 hour to reach 16 hours at 23 weeks (data from the University of KwaZulu-Natal).*

guides are in reasonable agreement with each other. Most advocate an intensity of about 60 lux for the first couple of days, followed by a gradual reduction to between 5 and 10 lux by 1 to 3 weeks. The intensity is then held at this level until 20 weeks, after which it is increased to between 15 and 40 lux for the duration of the laying period.

The similarity of the responses of broiler breeders and turkeys to daylength, no doubt because both species exhibit photorefractoriness, may indicate that the light intensity recommendations for turkeys can also be used for broiler breeders. Studies of the effect of light intensity on egg production suggest the ideal light intensity for turkeys is between 30 and 50 lux (Figure 5.14, page 88). Coincidentally, this is in good agreement with the recommendations from most of the primary breeders. In the absence of any evidence to indicate otherwise, it seems sensible, therefore, to follow the breeders' guides, especially as a high proportion of broiler breeders are kept on litter, where the light intensity needs to be bright enough to dissuade hens from laying eggs on the floor.

Welfare regulations in the EU countries stipulate that the light for breeding birds must be bright enough to see clearly and to stimulate activity.

Twilight

Behavioural studies have shown that laying hens given an evening twilight period settle for the night in a more orderly fashion than when the lights are switched off abruptly. Although the provision of a twilight period is not included in the welfare regulations for breeding birds (but is required for laying hens), a 15-minute period of dim light (about 1 lux at the height of a bird's head) does seem appropriate in broiler breeder houses where it is common for the birds to spend the night on a raised slatted area.

Light colour and source
Light colour

In recent times, there have been claims that red compact fluorescent lamps beneficially influence egg production and shell quality. However, these claims do not appear to be supported by any scientific research. The report that red light was more stimulatory than other colours of light (pages 11 and 12) seems to have been misinterpreted. The original research was conducted using truly monochromatic light sources, but, as shown in Figure 10.2 and Table 10.4, coloured fluorescent lamps have a very broad spectrum output. Furthermore, it should be remembered that white light includes red, and so there is no need to replace white lamps to provide the birds with red light, especially where incandescent lamps are used. In fact, the spectral output data in Table 10.6 show that a white incandescent lamp produces almost four times as much red light as a typical 'red' compact fluorescent lamp.

Light source

The type of lamp has no known influence on egg production, shell quality, fertility or liveability, and so, as discussed in the rearing section, the choice among incandescent, fluorescent or sodium vapour lamps will depend largely on economics and availability. However, some research has suggested that ultraviolet light might have a beneficial influence on fertility in broiler breeders because it is implicated in sexual recognition between males and females. Studies of the response of broiler breeders to a recently developed 'bird-lamp' and its effect on performance and behaviour are being conducted at the University of KwaZulu-Natal in South Africa. The bird-lamp emits UV-A as well as white light, and its output matches the spectral composition of sun light more closely than other fluorescent lamps.

Table 10.6 *Spectral output (%) from incandescent white and red compact fluorescent lamps between 350 and 780 nm.*

Colour	Incandescent white lamp	Fluorescent red lamp
UV-A	0.2	0.0
Violet	1.0	6.9
Blue	3.6	5.4
Green	8.7	6.0
Yellow	7.6	11.9
Orange	8.4	51.6
Red	70.5	18.2

11 LIGHTING FOR BROILERS

In the early days of the broiler industry, birds were given either continuous illumination or, more often, 23 hours of light and a 1-hour dark period to acclimatise them to darkness in case of a power failure. Although this is still typical in some parts of the world, concerns for bird welfare and the need to conserve energy have prompted the use of shorter daylengths or alternative lighting programmes. In contrast to lighting for replacement pullets, there is no concern for the effect of lighting on sexual maturity in broilers, simply its influence on feed intake, growth, feed conversion efficiency and welfare. This chapter describes various lighting programmes, including those that optimise feed conversion efficiency without compromising growth, improve liveability and minimise leg problems and ascites.

Constant daylengths

Feed intake and growth

Broilers were originally given continuous or near-continuous light to maximise feed intake and body weight gain. Body weights for birds given continuous illumination were about 5% higher than those of birds maintained on 12-hour days and 10% higher than those given 8-hour days. Modern strains of broilers seem to respond differently from earlier genotypes to daylength, especially to daylengths that are 12 hours or longer. Whereas feed intake and growth to 21 days are still positively correlated with daylength, negative correlations after 21 days result in feed intake to, and body weight at, 49 days being similar for all daylengths from 12 hours. However, broilers given 8-hour daylengths still have lower feed intakes and slower growth than birds given 12 or more hours of light. The data in Figure 11.1 show the contrasting influences of daylength on growth up to and after 21 days of age. The literature does not report data for broilers given daylengths of 9, 10 or 11 hours, but the linear effect of daylength on growth between 8 and 23 hours suggests that body weights at 21 days would be intermediate between those for 8 and 12 hours. However, the effect of 9, 10 and 11-hour daylengths after 21 days is unknown, and so the minimum daylength at which body weight at 49 days is not significantly affected by the duration of light is unknown. However, it is unlikely that these shorter days would be considered for use in commercial broiler-growing operations.

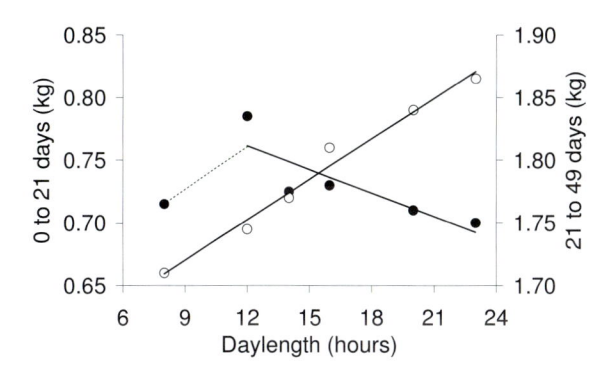

Figure 11.1 *Body weight gain to 21 days (○) and between 21 and 49 days (●) for broilers maintained on daylengths of between 8 and 23 hours (data from Tables 3.3 and 3.4, page 39).*

Mortality and culling

Total mortality (including culls), and the incidences of Sudden Death Syndrome and leg problems increase linearly with daylength. The data in Figure 11.2 were obtained from two experiments and, despite the total mortality being higher than that typically recorded for commercial crops of broilers to 49 days, clearly show the adverse effect that longer daylengths have on liveability. With no benefits for growth or feed conversion efficiency, but higher mortality, more leg problems and larger electricity costs, there is no reason to give broilers more than 12 hours light per day.

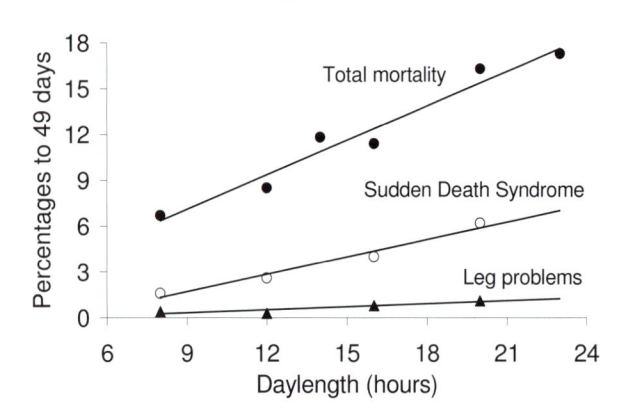

Figure 11.2 *Percentage total losses, Sudden Death Syndrome and leg problems to 49 days for male and as-hatched broilers maintained on daylengths of between 8 and 23 hours (data from Table 3.3, page 39).*

Changing daylengths

Broilers that have been given 6-hour days for the first 21 days and 23 hours thereafter have similar body weights at 42 days to birds grown throughout on 23 hours. However, they convert feed more efficiently and have better liveability, with fewer incidences of Sudden Death Syndrome and less leg problems (see Table 3.4, page 39). The improvement in health is attributed to slower initial growth. Although compensatory growth after the birds have been transferred to 23-hour days results in them having similar body weights at depletion to birds grown on long daylengths from day old, lower body weights during the first 3 or 4 weeks mean that loads on leg joints are reduced; skeletal development is related more to age than to body weight.

Some of the accelerated growth after the birds have been transferred to 23-hour days, which contrasts with the slower growth during this period when birds are maintained on long days from day one, may be a response to an increase in testosterone induced by the increase in daylength.

A further advantage for the initial use of 6-hour days is the saving of 119 hours of electricity per week compared with birds given 23-hour days. However, this programme may not be financially viable when birds are killed before 42 days of age, despite the improvements in bird health, because there will likely be insufficient time for the birds to make up the deficit in body weight that exists at 21 days.

Intermittent lighting

Short-cycle intermittent lighting programmes, such as repeating cycles of 1 hour of light and 2 or 3 hours of darkness, significantly improve feed conversion efficiency by increasing body weight whilst, at the same time, decreasing feed intake (Figure 11.3). However, it is important that the birds are given an initial 7 days of conventional 23-hour daylengths before they are transferred to intermittent lighting. The benefits are mainly achieved by a reduction in activity-related energy expenditure and, to a lesser degree, by an improvement in the metabolisability of the diet (see Table 4.4, page 62). It has also been suggested that intermittent systems encourage the birds to 'meal-feed' rather than 'nibble' and that this leads to a more efficient use of nutrients.

Feed intake and body weight gain are temporarily depressed when chicks are transferred abruptly to short-cycle intermittent lighting and so it may not be wise to use this form of lighting when birds are killed before 42 days because there may be insufficient time for the birds to regain lost growth. This is particularly so for females, who may take until 56 days to compensate.

Intermittently illuminated broilers also have an improved immune response and, because of reductions in oxygen consumption and heat production during the dark periods, fewer deaths due to ascites and hydropericardium. There are also substantial savings in electricity.

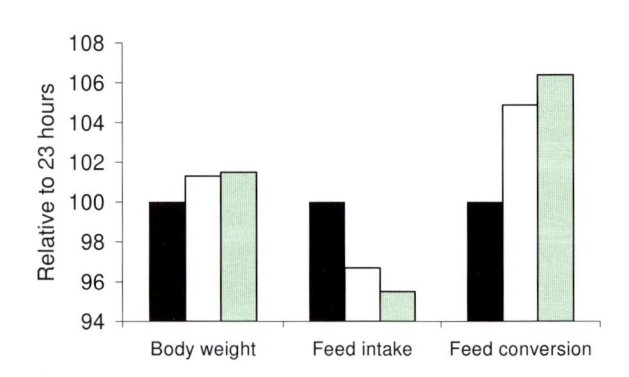

Figure 11.3 *Body weight, feed intake and feed conversion efficiency for as-hatched broilers given 1L:2D (white bars) or 1L:3D (grey bars) light-dark cycles from 7 days relative to birds maintained on 23-hour days throughout the growing period (black bars) (data from Table 4.3, page 62).*

Light intensity

Surprisingly, there is a small but significant depression of growth and a reduction in feed intake when broilers are given a brighter light intensity (see Figure 5.10, page 86). However, in practical terms, light intensity between 1 and 200 lux has a very small effect on broiler performance traits.

There is a European Union welfare directive that requires birds to see and be seen clearly and light intensity to be bright enough to enable the birds to be satisfactorily inspected. It therefore seems apposite to recommend an intensity of between 5 and 10 lux at bird level.

Evening twilight

Behavioural studies have shown that broilers on 12 or 14-hour days are better able to predict the oncoming darkness if they are given a period of decreasing light intensity than when the lights go out abruptly. The evening twilight stimulates an increase in feed intake, which, together with a reduction in energy expenditure, results in heavier body weights and improved feed conversion efficiency.

The provision of an evening twilight has been reported to be beneficial to the welfare of laying hens by allowing them to settle for the night in a more orderly fashion, and there is no reason to assume that a similar situation would not occur in broilers. It seems prudent, therefore, on both performance and welfare grounds, to provide a 15 to 30-minute period of dim light (about 1 lux) at the end of the main light period. This can be simply arranged by having a secondary lighting circuit of low wattage (7 or 8 W incandescent) lamps located at double the normal spacing of the main lights.

Light colour and source

Notwithstanding that research with truly monochromatic light has clearly demonstrated superior growth for broilers (and turkeys) given blue or green light rather than white or red (see Figure 6.2, page 96), growth in trials in which commercial coloured lights have been used has rarely been affected. This is because there is not really an improvement in growth under blue and green light but a depression of growth under red or white light. White light includes red, and most commercial coloured lamps emit light across the whole visible spectrum, including some red (Table 10.4, page 140).

Research with monochromatic light sources has suggested that broilers should be started on green light and then transferred to blue light at 10 days, but the results were equivocal, and it was only a suggestion that there might be some benefit by changing light colour during the growing period (see page 96). Furthermore, it must be emphasised that these experiments were conducted using truly monochromatic light sources and not commercial coloured lamps. The benefit of using commercial coloured lamps has yet to be investigated in a scientifically conducted experiment. Commercial coloured lamps are vigorously marketed around the world, but a producer should first enquire how monochromatic the lamps are, and secondly request evidence of their proven efficacy at improving growth, a clear case of *Caveat emptor* - buyer beware!

There have been many studies of the effect of light source on broiler performance, but none has unequivocally demonstrated any benefit for one light source over another. This is partly because the light source has frequently been confounded with differences in light intensity or colour. Indeed, similar body weights have been recorded for broilers reared under lighting conditions as diverse as 15W incandescent at 2 lux to 40W fluorescent tubes at 20 lux. It is therefore concluded that the choice of lighting will depend on capital outlay, running costs and ability to dim rather than on a lamp's influence on performance or health.

Which lighting programme to use?

It is clear that, when broilers are grown to at least 42 days of age, the provision of 6-hour daylengths for the initial 21 days and 23-hour days thereafter or short-cycle intermittent lighting from 7 days have advantages in terms of feed conversion efficiency and bird health over conventional broiler lighting regimes, especially those that provide continuous or almost continuous illumination. Estimates of the relative influence of each type of programme are given in Table 11.1, and these show that very long daylengths result in performance that is inferior to all other types of lighting.

Table 11.1 *Characteristics and influences of various lighting systems for growing broilers to at least 42 days relative to 23 hour-days or continuous illumination.*

Characteristic or trait	Conventional daylengths 12 hour	Conventional daylengths 23 hours or continuous	6 hours to 23 hours at 21 days	Short-cycle intermittent
Feed intake	99	100	98	96
Body weight	99	100	100	101
Feed conversion efficiency	100	100	102	106
Liveability	111	100	105	100
Leg problems	same	-	reduced	same
Ascites and hydropericardium	same	-	same	reduced
Electricity savings	yes	-	yes	yes
Lightproof housing	preferred	-	yes	preferred

12 LIGHTING FOR BREEDING TURKEYS

Turkeys are seasonal-breeding birds and so, without doubt, one of the most important functions of the lighting regime is the dissipation of photorefractoriness so that birds can be stimulated into sexual development at an appropriate age. It is important that males (variously called stags or toms) should be fertile during the pre-laying period and so males and females should be given different lighting treatments, both in daylength and intensity. This chapter describes typical programmes for matching sexual maturity in the sexes, and suggests areas where further studies might reveal better ways of achieving this aim.

• Lighting for female breeders •

Rearing period

The main objective of a lighting programme for turkey females during the rearing period is the provision of short days to dissipate photorefractoriness, a condition in which the turkey is unable to satisfactorily respond to a transfer to long days (see pages 16-18). Turkey females do not need to be kept on short days for the whole of the rearing period: the number of weeks of short days required to dissipate photorefractoriness depends on the daylength used (see page 42). Turkeys do not exhibit an absolute form of photorefractoriness, because they can mature spontaneously even when held on constant long days. This is apparent when females that have been inadvertently included in a flock of commercial males through mis-sexing, or intentionally kept to 22 to 24 weeks of age to suit market demands, start squatting in the final weeks of the growing period.

During the first 3 or 4 months there is no need to be too concerned about the light-tightness of the accommodation; indeed, birds are frequently given natural lighting at this stage of rearing. However, it is vital that houses are completely lightproof during the pre-laying period if unwanted sexual development is to be avoided.

Initial 4 months

It is normal practice to give turkey breeders an initial 1 or 2 days of continuous light at an intensity of about 100 lux to allow them to find feed and water, and to settle into the house. This is followed by a couple of days of 23 hours light before an abrupt transfer at 4 days to natural lighting or to a 14-hour day at an intensity of 30 to 50 lux (as measured at bird-head height). It is important to give these long days for the first 3 to 4 months to ensure that the birds do not dissipate photorefractoriness too quickly; otherwise there is a risk that the birds will start to mature while they are still on short days and at too young an age.

Pre-lay daylengths

At between 3 and 4 months of age, the daylength should be reduced by 1 hour per week to reach 6 or 7 hours by 18-21 weeks (Table 12.1), but the step-down lighting is probably given more for management than for photoperiodic reasons. One breeding company even suggests there may be an occasional need to go down to 5 hours. This variation in lighting advice from the primary breeding companies might reflect specific lighting requirements for particular genotypes, but this seems unlikely. In most experiments, photorefractoriness has been dissipated by providing 8-hour daylengths and, in some countries, an 8-hour daily light period is stipulated as a minimum in welfare regulations. The reason for the recommendation to give 6 or 7-hour daylengths is pragmatic; occasionally some flocks start to mature (as evidenced by squatting) when given 8 hours of light, and the provision of a shorter daylength appears to prevent or at least reduce the incidence of precocious sexual development.

Chapter 12 Lighting for breeding turkeys

Table 12.1 *Typical lighting programmes for male and female turkey breeders, with pre-lay daylengths for females highlighted in bold type.*

	British United Turkeys (closed housing)				Hybrid Turkeys (closed or curtain-sided housing)			
	Females (hens)		Males (toms)		Females (hens)		Males (toms)	
Age	(hours)	(lux)	(hours)	(lux)	(hours)	(lux)	(hours)	(lux)
1 day	23-24	100	23-24	100	24	70	24	70
2-4 days	23	100	23	100	23	70	23	70
4 days to 12 weeks	14	30	14	30	14/NAT [1]	50	12	40
12 weeks	**13**	35	13	29	14/NAT	50	12	40
13 weeks	**12**	35	12	28	14/NAT	50	12	40
14 weeks	**11**	40	10	27	14/NAT	50	12	40
15 weeks	**10**	40	10	26	14/NAT	50	12	40
16 weeks	**9**	45	10	25	14/NAT	50	12	40
17 weeks	**8**	45	10	25	**8 (10)** [2]	50	12	40
18 weeks	**6 or 7**	50-60	10	25	**8 (10)**	50	12	40
19 weeks	**6 or 7**	50-60	10	25	**8 (9)**	50	12	40
20 weeks	**6 or 7**	50-60	10	25	**8 (9)**	50	12	40
21 weeks	**6 or 7**	50-60	10	25	**7 (8)**	50	13	40
22 weeks	**6 or 7**	50-60	10	25	**7 (8)**	50	13	40
23 weeks	**6 or 7**	50-60	10	25	**7**	50	13	40
24 weeks	**6 or 7**	50-60	14	20-25	**7**	50	13	40
25 weeks	**6 or 7**	50-60	14	20-25	**7**	50	14	30
26 weeks	**6 or 7**	50-60	14	20-25	**7**	50	14	30
27 weeks	**6 or 7**	50-60	14	20-25	**6 or 5**	50	14	30
28 weeks	**6 or 7**	50-60	14	20-25	**6 or 5**	50	14	30
29 weeks	14	100	14	20-25	**6 or 5**	50	14	30
30 weeks	14	100	14	20-25	14/NAT	100	14	30
31-33 weeks	14	100	14	20-25	14/NAT	100	14	30
34 weeks	14.5	100	14	20-25	14/NAT	100	14	30
35 weeks	15	100	14	20-25	14/NAT	100	14	30
36-38 weeks	15	100	14	20-25	15/NAT	100	14	30
39-44 weeks	15.5	100	14	20-25	15/NAT	100	14	30
45-50 weeks	16	100	14	20-25	15/NAT	100	14	30
> 51 weeks	16.5	100	14	20.25	15/NAT	100	14	30

[1] NAT natural lighting
[2] Daylengths in brackets are recommendations for flocks transferred to long days in the autumn

Information in this table has been obtained from management guides produced by the following primary breeding companies (current in 2005).

British United Turkeys Ltd
Broughton
Chester
United Kingdom

Hybrid Turkeys
Kitchener
Ontario
Canada

In periods of high temperature and when birds are kept in non-controlled environment accommodation it is recommended to provide longer daylengths during the step-down period to alleviate heat stress (Table 12.1).

Turkeys will regard these short days during the pre-lay period as being photosexually neutral and may start sexual maturation spontaneously if photostimulation is delayed. Another possible cause of precocity in birds on short days is inadvertent exposure, for whatever reason, to a long day. This could be a time switch malfunction, an incorrect setting or even a deliberate over-riding of the clock to carry out some emergency work in the house. Accidents happen, but the light period should never knowingly be extended if undesirable early sexual development is to be avoided. Commercial experience has shown that once sexual development has started, as indicated by a proportion of birds squatting, there is little that can be done to halt it. In such circumstances it is probably better to be bold and transfer the flock to a 14-hour day, even though the resulting advance in maturity will result in inferior rates of lay (Figure 3.32, page 43) and the production of smaller eggs, rather than to risk creating an uneven flock.

Pre-lay light intensity

The main reason for recommending the use of a relatively bright light intensity (50 to 60 lux) during the pre-lay period seems to be to provide a bigger contrast with any extraneous light in houses that are inadequately light-proofed and, hopefully, to encourage the bird to adopt only the 6 or 7 hours of artificial light as its day. However, in an experiment in which turkeys were given 6 hours of light at 650 lux and 18 hours of darkness or dim light at various light intensities between 0.5 and 5 lux, any dim light given during the 'night' had a dire effect on subsequent egg production (Figure 12.1).

The marked difference between the production of hens given 18 hours darkness and those given dim light shows that it is essential to thoroughly lightproof turkey accommodation in the pre-breeding stage.

Some breeding operations suggest that the brighter light intensity during the rearing period also helps to promote activity when the birds are on very short days.

It seems clear that more research is required to establish the most appropriate light intensity to use during this phase of the rearing period, and to quantify the effect of light leakage that creates a bright-dim-dark rather than a bright-dim cycle as used in the above experiment. It should be appreciated that houses need to be very well light-proofed to give a light intensity that is below 0.5 lux during the periods coinciding with natural daylight.

Photostimulation age

The most important decision when photostimulating turkey females is the age at which they are transferred to long days. Data in Figure 12.2 show that between 20 and 32 weeks the average rate of lay increases progressively by about 1% for each 10-day delay in the age at

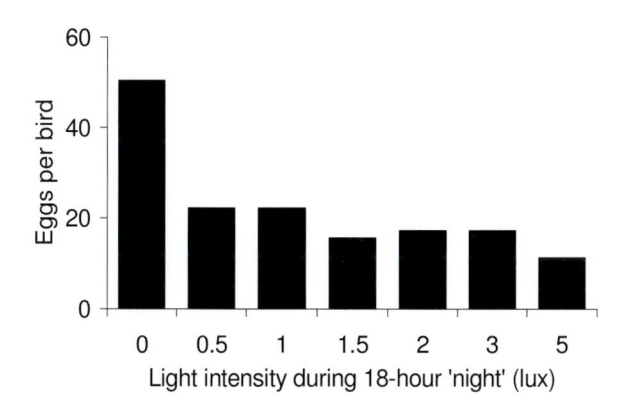

Figure 12.1 *Effect of various pre-laying intensities of light during the 18-hour 'night' on subsequent egg production in turkeys given a 6-hour day at 650 lux.*

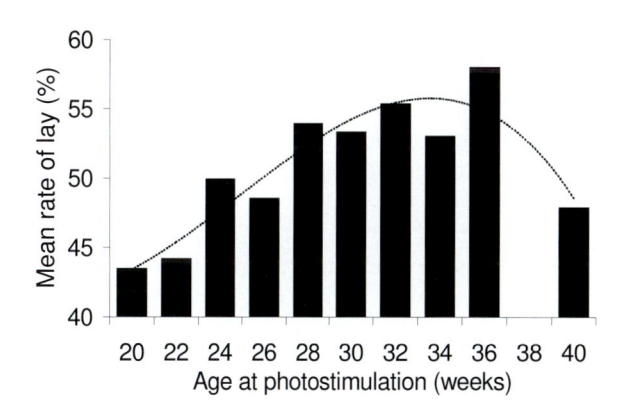

Figure 12.2 *Effect of age at transfer to a stimulatory photoperiod (often 14 h) on mean rate of lay for a fixed period from photostimulation.*

which the birds are photostimulated, but then decreases when the transfer to long days is not made until after 34 weeks of age. In the turkey industry, most birds will be photostimulated at about 30 weeks (Table 12.1) so that egg production starts at around 32 weeks. Although the regression curve in Figure 12.2 suggests that maximum egg production would be achieved by delaying photostimulation until 34 weeks of age, this is likely to be uneconomic because the improvement in rate of lay translates into little more than 3 extra eggs over a 24-week production period, and the extended rearing period would require about 15% more feed to get the birds to sexual maturity.

Daylength in the laying period
It is usual to give turkeys an abrupt transfer to long days (often 14 hours), which ensures a compact start to egg production. When a flock is given a series of increments in daylength, it takes longer to reach peak rate of lay because the spread of individual maturities is protracted and this results in no better egg production over the laying cycle. The delay in age at 50% lay does give some increase in egg weight and, as a consequence, poult size early in the laying period, but this advantage is more than cancelled out by the staggered entry of individual birds into production, which creates problems with the timing of artificial insemination in the early stages of the laying period.

However, should a producer elect to use a step-up programme, the initial increment must be large enough to ensure that sexual development is triggered in all birds.

Egg production is maximised by going to a 12 or 13-hour day (see Figure 3.29, page 42), but this can only be achieved in controlled environment housing. Many turkey flocks are kept in non-controlled environment accommodation and thus experience daylengths that are longer than 13 hours in mid summer at high latitudes. If these birds were given only 12 or 13 hours of light after midsummer, they would suffer a decrease in daylength after the longest natural day, and this would adversely affect egg production.

It is common practice to give further increases beyond 14 hours from about 35 or 36 weeks of age to reach a maximum of 15, 16 or even 17 hours. Whilst this may be important in non-controlled environment housing to ensure that there is no decrease in daylength during the laying period, this is probably counter productive in lightproof housing because longer daylengths simply accelerate the onset of adult photorefractoriness (see pages 16-18).

For the more adventurous producers with lightproof houses, intermittent lighting programmes can be used to save electricity. However, it is recommended that the flock is given 4 weeks of conventional 14-hour days before the introduction of intermittent lighting so that the hens learn to use the nest boxes. If the birds are transferred to an intermittent programme immediately after the pre-lay short-days, floor-laying can be a problem. Details of this type of lighting are given on pages 63-64.

Light intensity during the laying period
Typically, producers are advised to use a light intensity of 100 lux during the laying period, although experimental findings indicate that egg production will be maximised by about 50 or 60 lux (see Figure 5.14, page 88). There seems to be no good photobiological reason why higher intensities are used.

• Lighting for male breeders •

It is surprising, bearing in mind that turkeys are seasonal breeders, that it is not recommended to give males a period of short days during rearing. Primary breeding companies state that males are slower maturing than females, but the females will have been given a period of short days, thus making them photoresponsive at a younger age than the males. However, it is doubtful that this difference in lighting philosophy for males and females is a consequence of photobiological reasoning, more likely a case of this is how things have always been done. Traditionally males were kept in simple open-sided pole-barns, and so could not be given daylengths that were shorter than those prevailing naturally. Thus, to minimise the consequences of the changes in natural daylength, they were reared on relatively long days. This method of lighting still prevails, except that males are now reared on 10 to 12-hour days, if they are kept in closed housing, and then transferred to long days 4 to 6 weeks before the females are photostimulated (Table 12.1). Alternatively, they may be allowed natural lighting in open or curtain-sided houses.

Daylength

During the initial 3 or 4 days, males, like females, are given either continuous or near-continuous illumination to get them off to a good start before abruptly changing them to 10 to 12-hour daylengths for about 5 months.

Commonly, males in lightproof buildings are transferred abruptly to 14-hour days at 24 weeks or given a step-up programme from 20 weeks to reach 14 hours by 24 weeks, and maintained on this daylength thereafter. However, in poorly light-proofed houses, it is suggested that the birds be given 30-minute increments every 2 weeks from 30 weeks to a 16 or 17-hour maximum to ensure that the birds do not experience a decrease in daylength. Males in open or curtain-sided housing are given supplementary artificial light so that daylength is never less than 14 hours.

Light intensity

Males have a tendency to be aggressive, and so it is usual to keep them at a lower light intensity than that used for females. In lightproof houses this is usually between 20 and 30 lux at bird-head height. In open-sided housing, it is recommended to use plastic woven-mesh screening to minimise direct sunlight, although it is important that this is regularly cleaned so as not to restrict ventilation.

Keeping males, be they first or second-year birds, at lower light intensities to control aggression does not appear to compromise any aspect of fertility, with illuminance as diverse as 5 and 40 lux having no effect on semen volume, sperm concentration, sperm numbers per ejaculate or subsequent hatchability.

Light colour and source

There is no unequivocal evidence to suggest that the use of one type of lamp results in any better performance in turkey breeders than any other type. Thus the choice of light source depends on availability, capital outlay, running costs and the ability to dim the lights using conventional voltage reduction equipment.

Despite the claims of some lamp manufacturers, there are no proven benefits from the use of coloured lights, and so there is no need to use other than white light for illuminating turkeys, in either the rearing or the laying period.

• Recycling turkeys for a second production cycle •

Breeding turkeys are sometimes kept for more than one production cycle. Although turkey hens appear to spontaneously start another cycle of egg laying, it is not practical or economic to leave the recycling to the birds themselves, because production will fall to almost zero before a large proportion of the flock rests for about 2 months (see Figure 2.10, page 18). It is therefore more efficient to reduce the daylength to 6, 7 or 8 hours for about 2 months after the birds have been in production for 24-26 weeks (when egg production will have fallen to about 60%) and then to abruptly transfer them back to 14-hour days. The objective of the short days is to dissipate adult photorefractoriness, the condition that causes a cessation of egg laying, so that the hens can again be responsive to a stimulatory daylength. It is important when choosing the short day that a producer is aware of any welfare regulations that might stipulate minimum daylengths for breeding turkeys.

Experiments have shown that recycling can also be achieved by dramatically reducing the light intensity whilst keeping the hens on long days. However, the intensity needs to be so low, with a threshold somewhere between 0.5 and 2 lux (see Table 5.5, page 88), that to achieve this level of lighting the houses have to be completely lightproof. It is much simpler to reduce the daylength to between 6 and 8 hours to regain photosensitivity.

• Ovarian carcinoma •

Turkey hens exposed to long days can develop spontaneous ovarian tumours, especially when they are kept for more than one laying cycle. Amazingly, during the recycling process, the tumours seem to regress completely within 4 or 5 weeks of the birds being put onto short days, but unfortunately they will regenerate within 5 to 6 weeks of the birds being transferred back to long days.

• Polyovarian Follicle and Polycystic Ovarian Follicle Syndrome •

Recently, a new disease condition called Polyovarian Follicle and Polycystic Ovarian Follicle Syndrome has been observed in breeding turkeys transferred from 6-hour days to continuous illumination at 30 weeks. The condition was characterised by hens ceasing to ovulate after only 2 to 3 weeks of becoming sexually mature, but still recruiting follicles into the ovarian hierarchy. Some of these follicles continued to grow but became cystic, whilst affected hens occasionally retained a shelled egg in the oviduct for many days or even weeks. It has been suggested that the condition is caused by a persistently high phosphorus concentration in the blood stream, and that this inhibits the pre-ovulatory rise in luteinizing hormone that is necessary for a follicle to be ovulated. Simultaneously, however, normal concentrations of follicle stimulating hormone and oestradiol continue to stimulate the recruitment of follicles into the ovarian hierarchy.

The incidence of the condition is substantially higher in turkeys transferred to continuous illumination than in those given only 14-hour daylengths, especially when the birds have been photostimulated at or before 32 weeks of age (see Figure 7.6, page 106).

Whilst it is appreciated that commercial flocks of breeding turkeys are unlikely to be transferred to continuous illumination, the significant difference in the incidence of the condition in birds given continuous light and those on 14-hour days (Figure 7.6, page 106) does suggest an association with excessive illumination. It would therefore be prudent to monitor flocks that are given very long daylengths, such as 16 or 17 hours, and to note that the condition can still occur in birds that are given only 14 hours light.

13 LIGHTING FOR GROWING TURKEYS

Lighting for growing turkeys used to be the simple provision of a 23-hour day to maximise feed intake and growth. However, the importance of an adequate dark period for health and skeletal integrity and the fact that daylength has a relatively small effect on feed intake means that turkeys are now given shorter days. This chapter discusses all facets of lighting for growing turkeys, including intermittent programmes, the influence of light intensity and colour on growth and the different responses of males and female to daylength.

• Conventional daylengths •

Although body weight during the first 6 weeks tends to be related to daylength, turkeys given 12 or less hours of light learn to eat in the dark (Figure 13.1) and, by about 15 weeks, daily feed intake and body weight are similar for most daylengths. However, from 15 or 16 weeks of age, males become sexually responsive to light, and those given stimulatory daylengths ($\geq$ 12 hours) increase their feed intake and convert feed more efficiently to produce heavier body weights by 20 weeks (Figure 13.2 and Table 3.7, page 41).

Data in Table 3.7 show that male turkeys given a programme of 16 hours light and 8 hours of darkness had similar body weights but better feed conversion efficiency to 20 weeks than birds maintained on 12 hours light, and heavier body weights but slightly poorer feed conversion efficiency than birds given 23 hours light. However, during the period of sexual development, the birds given 16 hours light had a higher feed intake, faster growth and more efficient feed conversion than the 23-hour birds, and so 16 hours is likely to be more profitable than 23 hours if birds are kept beyond 20 weeks.

In addition to these growth benefits, the 8-hour period of darkness for birds on 16-hour days allows them to release more of the nocturnally synthesised hormone that controls calcium mobilisation, thus optimising skeletal development and, as a consequence, reducing the number of leg problems compared with a flock given 23 hours of light and only 1 hour of darkness. The shorter period of illumination will also result in lower mortality in turkeys that have not been beak-trimmed, de-toed or de-snooded (see Table 3.8, page 41). In time, it is likely that welfare regulations will make the provision of an adequate period of darkness compulsory, at

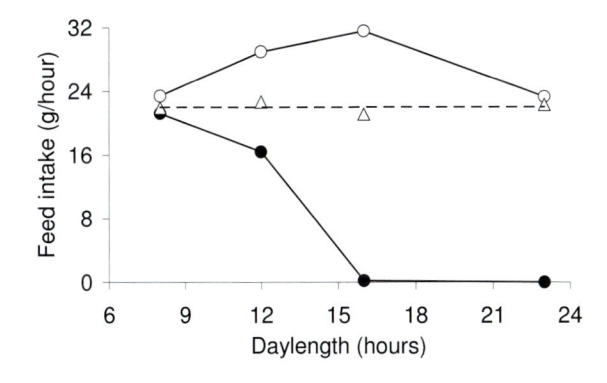

Figure 13.1 *Mean hourly feed intake during the light (○), darkness (●) and daily mean (∆, broken regression line) at 14 weeks for male turkeys given an 8, 12, 16 or 23-hour daylength from 2 days of age (data from the University of Bristol).*

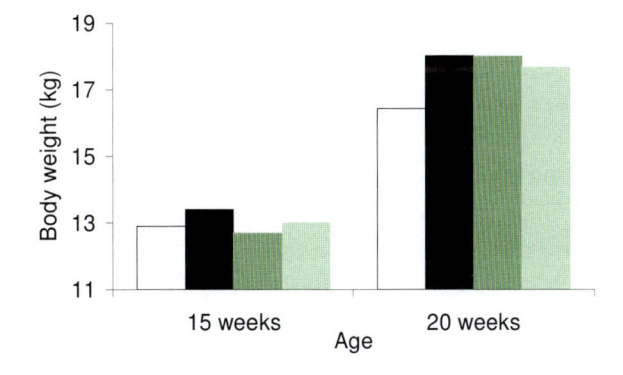

Figure 13.2 *Body weights at 15 and 20 weeks for male turkeys given 8 (white bars), 12 (black bars), 16 (dark grey bars) or 23-hour (light grey bars) days (data from the University of Bristol).*

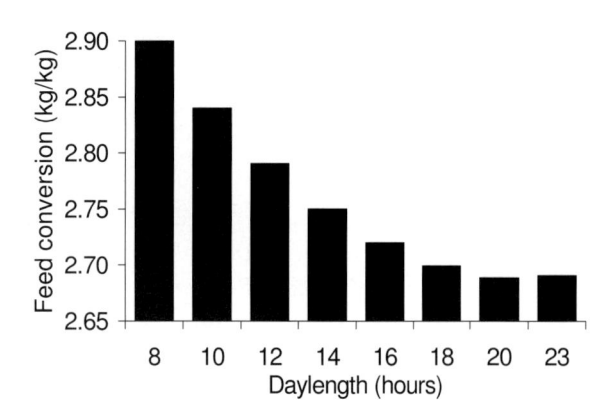

Figure 13.3 *The effect of daylength on feed conversion efficiency to 20 weeks in male turkeys. Regressed data from Figure 3.29 (page 41).*

least in European Union (EU) countries. Currently, UK welfare codes for turkeys advise the use of a minimum of 8 hours of light for birds not exposed to natural lighting, which is the original recommendation for laying hens. That has now been changed to state a minimum dark phase of about 8 hours and it seems probable that the turkey codes will be similarly amended.

Commercial programmes

Although in small-group experiments there are generally no problems in using daylengths of any duration from day old, it is prudent in the more competitive environment of large-scale commercial turkey growing operations to give 1 or 2 days of continuous illumination, so that the birds easily find feed and water. This is then followed by a series of decreases to reach the selected daylength by the end of the first week.

Although the data in Figure 13.2 show that there is little difference in the body weight of males grown on 12, 16 or 23-hour daylengths, and unpublished findings from one of the primary breeding companies revealed no significant difference between body weights for birds on 14 and 23-hour days, the information in Figure 13.3 shows that there is a marked improvement in feed conversion between 12 and 16 hours, which suggests that 16 hours is probably the optimal daylength for achieving the best overall performance in males.

Notwithstanding any obligation to comply with welfare regulations, it seems likely that fewer commercial turkeys will be grown on 23-hour daylengths in the future. Days of moderate length result in body weight and feed conversion comparable with that of turkeys given 23 hours, but the moderate daylength gives better liveability, fewer leg problems and reduced electricity costs. Furthermore, unpublished data from commercially sponsored research has indicated that there can also be lower incidences of Sternal Bursitis and Focal Ulcerative Dermatitis (breast buttons) when turkeys are given moderate (14 hours in this trial) rather than 23-hour daylengths. These two conditions can be major causes of downgrading in processing plants.

Specific female needs

There is no evidence to suggest that females need a different daylength from males. Commercial females are generally killed at younger ages than males and, as a consequence, do not show the androgen-induced accelerated growth that occurs during the last few weeks in males exposed to stimulatory light. However, oestrogen does influence growth and, if females are kept beyond about 18 weeks, it is likely that some individuals will respond to the light, undergo sexual development, start squatting and commence egg production.

Midnight feeding

Some breeding companies recommend the provision of a 1 or 2-hour period of light at around midnight to stimulate feed intake. The provision of a night-interruption photoperiod might increase overall intake in males on ≥ 16-hour daylengths because the data in Figure 13.1 show that these birds will consume very little during the dark. However, a midnight feeding period also creates an intermittent lighting regime, with the longer of the two dark periods being adopted as night while the other is linked to the two periods of light to form a very long subjective-day. For example, 16 light, 4 dark, 2 light and 2 dark will be interpreted as a 20-hour day (16+2+2) and a 4-hour night. This 20-hour subjective day will be less sexually stimulatory than a single 16-hour light period and, as a result, probably be associated with slightly slower growth during the final 4 or 5 weeks in males grown to 20 weeks, a period in which increased testosterone in the blood increase feed intake and improve feed conversion efficiency (see Figure 3.28 and Table 3.7, page 41). Another disadvantage of midnight feeding is that it is unlikely to optimise skeletal development and might therefore increase the incidence of leg problems. A final difficulty is that it is undoubtedly unacceptable to animal welfarists, and would be unusable if regulations specify an 8-hour minimum dark period.

Increasing daylengths

Some lighting programmes include a series of increments in daylength from about 15 weeks for males and from 13 weeks for females to reach a peak of 18 hours. Notwithstanding that field experience might indicate otherwise, it is difficult to understand what effect an increase in daylength will have on performance when the data in Table 3.7 (page 41) clearly show lower feed intakes and smaller weight gains for 23-hour compared with 16-hour days during the period of sexual development. This is because 23 hours is less stimulatory than 16 hours.

Naturally lit houses

During summer months, there is no need to give other than natural light during the daytime in open-sided houses, although it has been suggested that turkeys should be given 1 hour of artificial light at around midnight to encourage feeding (see Midnight feeding section above for comment). During the short natural days of winter, it is recommended to give artificial light before and after the natural day to create a 16-hour day, for example, artificial lighting 04.00 to 07.00 and 17.00 to 20.00 (actual lights-off in the morning and lights-on in the evening will depend on natural sunrise and sunset times).

Seasonal effect

Turkey growth in naturally illuminated units will vary with the season, even though artificial light may have been used to create a constant daylength. In experiments in which artificial lighting has been used to mimic seasonal changes in natural daylength, simulated autumn-hatched birds had heavier 16-week body weights and better feed conversion efficiency than simulated spring-hatched turkeys, irrespective of whether the trial was actually conducted in the spring or in the autumn. However, trials run in the spring to summer period produced heavier body weights and more efficient feed conversion than in birds grown during autumn and winter. These responses probably reflect the differences in seasonal temperatures in North Carolina (latitude 36°N) where the trials were conducted.

Changing daylengths and intensity

Various high-illuminance (20 lux) step-up and low-illuminance (2.5 lux) step-down lighting regimes have been studied to assess their effects on leg integrity. Generally, the incidence of leg problems has been reduced by step-up lighting due, possibly, to increased exercise, but the effects seem to be dependent upon using a brighter light intensity because no differences in leg health were observed at 2.5 lux.

Body weight and feed conversion responses have been variable, but, where there have been significant influences, they appear to have been a result of the effect that daylength had on feed intake. For example, when weekly increases or decreases between 10 and 16 hours were given from 4 to 16 weeks of age, hens on the step-down regime initially had larger feed intakes and faster growth than step-up birds, but their feed intake and body weight gain were inferior at the end of the trial when the daylength was shorter. Feed intakes and body-weight gains in the middle period were similar when the daylengths were more or less the same.

These equivocal findings make it difficult to recommend the use of step-up lighting as a means of improving leg strength in turkeys, especially as they appear to have been tested against 23-hour rather than 16-hour daylengths.

In male and female broilers, the use of short, 6-hour days for the first 3 weeks (or half life) has been proven to reduce leg problems and improve liveability, with no adverse effect on final body weight or feed conversion (see Table 3.4, page 39). However, such lighting regimes do not appear to have been tested in growing turkeys. Maybe an 8-hour daylength to about 10 or 12 weeks, followed by a transfer to 16 hours would have similar beneficial health effects for turkeys, especially as some of the compensatory growth in broilers has been attributed to photostimulated increases in sex hormones when the birds are transferred from short to long days.

• Intermittent lighting •

To gain full benefit from intermittent lighting, birds must be kept in lightproof houses. It is also important that they are given an initial 2 to 4 weeks on a conventional lighting regime before being transferred to intermittent lighting.

Short-cycle programmes

With the above provisos, turkeys grown on any short-cycle repeating programme have invariably had superior performance to conventionally illuminated birds, be the cycle 1L:2D (1 hour of light followed by 2 hours of darkness), 1L:3D, 2L:2D, 2L:4D or 4L:2D. Typically, the benefits for males shown in Figure 13.4 are greater than for females, because the intermittent schedules are sexually stimulatory and the older killing ages for males allow them more time to respond to the photostimulated rise in sex hormones and subsequent increase in feed intake. However, females given a short-cycle programme also perform better than those given conventional lighting. Most of the extra growth results from an increase in feed intake, though occasionally there will be a small improvement in feed conversion efficiency.

Unlike the responses of broilers to short-cycle lighting (see Table 4.4, page 62, and Figure 4.5, page 63), little is known about the effects that intermittent lighting has on energy expenditure and hormonal rhythms in turkeys, and so any contribution that these might make towards the improved performance is speculative.

Asymmetrical intermittent programmes

An asymmetrical intermittent programme is one in which the turkeys are given more than one light and dark period per solar day, but with one of the dark periods being longer than the other (or the rest). This permits the bird to interpret the longer(est) period of darkness as night and the remainder of the 24 hours as day.

This type of intermittent lighting also increases body weight compared with birds grown on 23-hour days and, as for short-cycle programmes, is primarily achieved through an increase in feed intake. However, the use of

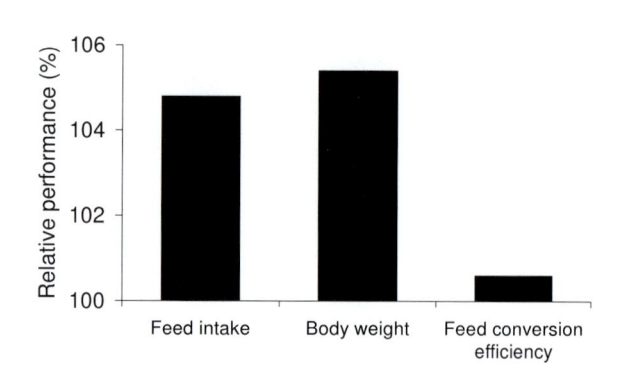

Figure 13.4 *Performance for male turkeys given short-cycle intermittent lighting from 4 weeks of age relative to birds maintained on 23-hour days.*

one particular programme in which male turkeys were given eight 1L:0.5D cycles followed by a 12-hour period of darkness resulted in precocious sexual development between 16 and 20 weeks and heavier body weights at 20 weeks than in birds given solid 8-hour daylengths. In contrast to other intermittent lighting experiments, the extra weight gain resulting from the use of this regime was achieved through an improvement in feed conversion efficiency compared with the short-day controls, presumably induced more by an increase in circulating testosterone than by an increase in feed intake.

Electricity savings

The total amount of illumination used in the various forms of intermittent lighting for growing turkeys ranges from 6 hours per day for a 1L:3D cycle to 16 hours for a 4L:2D cycle. Large electricity savings do not necessarily accrue from intermittent lighting, unless the comparison is made with 23-hour lighting; but that is not the optimal lighting programme.

Light intensity

Provided the light intensity during the first 1-2 weeks is at least 10 lux, higher intensities generally have minimal effect on growth, feed intake or feed conversion efficiency in turkeys. However, if poults are exposed to low intensity light (e.g. 1 lux) during the first 2 weeks, feed intake will be depressed, body weight gain reduced and mortality markedly increased.

Commercial recommendations

It is normal practice to beak-trim, de-snood and toe-clip commercial turkeys, and this allows the use of relatively bright illumination without adversely affecting the number of deaths or increasing the amount of culling of birds injured as a result of cannibalism.

A typical light intensity for the first 2 or 3 days is 35-50 lux at bird level. This is bright enough to ensure that all poults find feed and water. One North American company advocates that the light intensity is then progressively reduced to about 5 lux by the end of the first week, whilst others suggest a further decrease to reach 2 lux by 14 days.

There is no need to use a different light intensity for males and females, even though this would be possible in many cases because the two sexes are frequently grown separately.

It must be emphasised that these recommendations are for poults that have been beak-trimmed. In one trial in which male poults had been left intact and exposed to 8, 12, 16 or 23-hour daylengths at 1 or 10 lux from 1 day of age, mortality and the need to cull injured birds was so high at the brighter light intensity, that it was necessary, on welfare grounds, to reduce the intensity to 1 lux. Interestingly, the age at which it became necessary to dim the light to control cannibalism was correlated with daylength, occurring at 45, 73 and 75 days of age, respectively, for the birds on 23, 16 and 12 hours. Only the birds on 8-hour days could be maintained on 10 lux until 20 weeks. In another study, in which intact male turkeys were exposed to 12 hours of light at 5, 10, 36 or 70 lux, the incidence of tail and wing injuries to 5 weeks was also positively correlated with illuminance.

These findings pose problems for turkey producers in EU countries because welfare regulations are likely to be altered to recommend a minimum light intensity of 20 lux. Furthermore, there is the probability that beak-trimming, de-snooding and toe-clipping will also be prohibited in the near future.

Ultraviolet light

When producers are not allowed to beak trim birds and are also obliged to use a relatively high light intensity to comply with welfare regulations (e.g. 20 lux), one solution could be to provide supplemental UV-A light. In trials where turkeys were kept in groups of 100 and given UV-A and white light, it was possible to successfully rear intact males to 20 weeks at 70 lux with minimal cannibalistic behaviour and almost perfect feathering. However, this approach has not been tested commercially and, even in the research conditions, the full potential of UV-A lighting was only obtained when it was given in conjunction with environmental enrichment. An alternative may be to use the recently developed Arcadia bird-lamp. This lamp, which emits both UV-A (in a similar proportion to natural light) and white light, was originally designed for lighting exotic birds, many of which use ultraviolet reflective markings on their plumage to recognise mates. It is now being assessed for use in controlled environment poultry houses.

Light colour

Light colour

Distributors of coloured compact fluorescent lamps market their products extremely vigorously in many poultry industries around the world. They claim various benefits for the use of blue and green lamps for growing turkeys, but unfortunately can only produce performance information from field trials because there is little scientific evidence to back the claims other than results from experiments in which monochromatic light was used.

Growth

The faster growth of male turkeys to 16 weeks or female turkeys to 18 weeks when grown under blue as opposed to red light is without question (Figure 6.2, page 96). However, this research was conducted using monochromatic light, and emissions from commercial coloured lamps are far from monochromatic (see Table 10.4, page 140). Growth is not enhanced by blue or green light but is suppressed by red, and this suppression begins at about 565 nm, the point where green light changes to yellow. It is difficult, therefore, to accept that a commercial 'green' lamp that emits 50% green light, but also 25% in the yellow, orange and red range, can have the same influence on growth as a monochromatic lamp or an appropriately filtered white lamp. Typically, commercial blue lamps emit three quarters of their output in the blue and green wavelengths and only 3% as yellow, orange and red light, and are, as a result, more likely to have some beneficial influence on growth.

Once turkeys have reached an age from which they will start to develop sexually in response to a stimulatory daylength (15-16 weeks for males, 18-19 weeks for females), growth will be faster under red and white light than under blue. This is because longer wavelengths (that is red light or white light which, of course, contains red) are more stimulatory than short wavelengths of light (see pages 11-12). The improved growth observed in older turkeys under red light is a response to increased concentrations of sex hormones circulating in the blood. This is also the most likely explanation for the more efficient feed conversion during this period in turkeys lit with red as opposed to blue light. Consequently, if blue or green lamps are used in the hope of accelerating growth for the first part of the growing period, they should be replaced with white light after about 15 weeks, otherwise any early benefits will be lost. It appears that an objective assessment of the responses of poultry to these commercial lamps is required.

Behaviour

Male turkeys have been observed to be more docile, less active, show less sexually related activity and have fewer social interactions under blue light than under red or white. However, the docility and reduced activity, though possibly welcome from a social and welfare point of view, might have some undesirable consequences in other areas. For example, birds that spend more time sitting on the litter are more likely to develop breast blemishes than active birds, and this is liable to result in more carcasses being downgraded in the processing plant.

Mortality

With its various effects on sexual development, aggressive behaviour and activity, it is surprising that light colour, whether from a monochromatic or a commercial lamp, appears to have no effect on mortality.

• Light source •

Although growth and feed conversion efficiency have been better with one light source than with another in some experiments, the findings have been inconsistent, and it is concluded that the type of lamp is unimportant for growing either male or female turkeys.

There is also no evidence to show that one type of lamp has a greater effect on the number of leg problems or carcass downgrades than any other. There is one report that the proportions of wing and tail injuries under incandescent light were higher than under fluorescent light. However, it would be imprudent to discontinue the use of incandescent lamps on the strength of this one experiment, especially when the findings of another trial showed that mortality was higher under fluorescent than under incandescent light.

14 LIGHTING FOR WATERFOWL

A large proportion of the world's waterfowl, both ducks and geese, live in Asia and most are kept under natural lighting conditions. However, this is not to say that lighting is irrelevant for waterfowl. Indeed, recent research in Taiwan has shown how lighting can be used to make geese lay out of season when eggs can command up to three times the price of in-season eggs. This chapter discusses opportunities for lighting waterfowl rather than describing commonly used lighting programmes, as in the preceding five chapters. In Eastern Europe, there has been much research into lighting for geese, whilst in France there have been several studies using Muscovy or Barbarie ducks, which are kept intensively in lightproof housing. The interesting aspect of lighting for waterfowl is that some, like geese, are seasonal breeders, whilst others, like Khaki Campbell egg-laying ducks, are not.

• Muscovy (Barbarie) •

Daylength

Females

Muscovy females are not seasonal breeders, and so do not need a period of short days to make them responsive to a transfer to long days. Furthermore, they do not appear to be particularly responsive to photoperiod. In an experiment in which birds were given a step-down lighting programme from 24 hours to 14 hours at 26 weeks or maintained on 8.5-hour days between 4 and 16 weeks followed by weekly increases to also reach 14 hours at 26 weeks, ages at 50% egg production were similar. If such contrasting regimes had been given to either egg-type laying hens or broiler breeders the ages at sexual maturity would have been very different. However, the rate of change in age at sexual maturity for Muscovies transferred from 8 to 14 hours at various ages is similar to that for domestic fowl, with maturity occurring 1 day later for each 2-day delay in photostimulation (see page 44).

Males

In contrast to females, Muscovy males seem to exhibit some form of seasonal breeding. In a study in which birds were transferred from 6 to 16 hours light at 20 weeks of age, the birds initially responded, but within 24 weeks they underwent testicular regression, moulted and showed a decrease in plasma luteinizing hormone concentration. This termination of reproduction was followed later by a re-growth of the testes and an increase in plasma luteinizing hormone concentration without any change in daylength. This second spontaneous reproductive cycle, despite the birds being maintained on long days, is very similar to what has been observed in female turkeys (Figure 2.10, page 18) and other waterfowl. If the recrudescence is due to the same mechanism that occurs in female turkeys then a similar lighting programme to that used to recycle turkeys could probably be used to produce a second reproductive cycle in Muscovies. Whilst the birds might recycle spontaneously without a change in daylength, it is probably better to give the birds a period of short days and then return them to long days so that the breeding cycles occur in an organised way and supply fertile eggs when they are required.

Lighting programmes for parent stock

Although the evidence from studies of female Muscovies might suggest that daylength during the rearing period is unimportant, it will still be beneficial to rear them on short days so that they can be photostimulated at the right time to achieve the required age at sexual maturity. Commonly, Muscovies are recycled to achieve a second production cycle. At the end of the first cycle, birds are force-moulted by reducing the daily feed allocation and by returning them to short days for about 10 weeks. Following the moult, daylengths are changed back to 14 hours. A typical programme is given in Table 14.1.

Lighting programmes for commercial stock

Intermittent lighting is commonly used for growing commercial Muscovies when they are kept in lightproof accommodation. A typical programme provides continuous light for the first week, 18 hours light for the second, then a repeating 2 hours light:2 hours darkness or 1 hour light:3 hours of darkness regime through to depletion.

Table 14.1 *Typical lighting programme to achieve two production cycles in Muscovy (Barbarie) parent stock.*

Age	Daylength (hours)	Intensity
0 - 4 days	24	40-50 lux
5-7 days	22	
8-9 days	20	
10-11 days	18	
11-12 days	16	10 lux
13-14 days	14	
2-25 weeks	7	
26 weeks	9	
27 weeks	11	30-40 lux
28-51 weeks	14	
52 weeks	4	
53-54 weeks	6	10 lux
55-61 weeks	7	
62 weeks	9	30-40 lux
63-85 weeks	14	

Information (current in 2005) supplied by
SA Gourmaud Selection
St Andre-Treize-Voies
France

Light intensity, colour and source

There are no research findings available to help make definitive recommendations for light intensity, light colour or type of lighting. However, practical experience indicates that, following an initial few days at about 40 lux, a mean intensity of 10 lux at bird-head height during rearing and 30-40 lux after photo-stimulation will control undesirable behaviour whilst permitting optimal performance in parent stock. However, it may be necessary in some flocks to reduce the intensity during lay to 20 or 10 lux to calm the birds.

Light colour has little effect on reproductive performance in domestic fowl and turkeys, and so it is reasonable to assume that the use of white light will be perfectly acceptable for illuminating Muscovies.

Light source also has minimal influence on the performance of fowl and turkeys breeders, and so is unlikely to be of concern for Muscovies. Practical experience also shows that Muscovies perform equally well under fluorescent and incandescent light. The choice will therefore depend on the same factors as for other species of poultry, namely, availability, costs and the ability to dim the light if a lower intensity is required.

• Geese •

Daylength

Most breeds of geese are truly seasonal breeders, with sexual development in naturally illuminated birds occurring in early spring in temperate regions and in late autumn in subtropical parts of the world, when the daylength is between 10 and 12 hours. Egg production ends naturally after about 4 months, but always before the longest day.

Lightproof housing for rearing

It is recommended that geese be given a lighting programme similar to that used for rearing female turkey breeders. The birds can be exposed to natural lighting until 3 months before the desired start of egg production, at which stage they should be changed to 7-hour daylengths for a period of 8 weeks. This will make the geese responsive to a transfer to stimulatory daylengths. Photostimulation can be achieved by transferring the birds to a 14-hour day or returning them to natural lighting.

Sexual maturity will be delayed if geese are not given a period of short days to dissipate photorefractoriness, that is, the birds will take longer to become sensitive to a stimulatory daylength (see section on photorefractoriness, pages 16-18). Additionally, geese that are kept on long days will have inferior egg production, as shown in Figure 3.35 (page 45).

Laying period

Where it is possible to keep geese in lightproof housing, the season can be markedly extended by limiting the daylength to no more than 10 or 11 hours. Egg production has also been improved in geese that have completed one laying cycle under natural light by transferring them to 7-hour days for 8 weeks followed by a step-up programme of 30-minute increments to reach 10 hours after 6 weeks. The birds are then held on 10-hour days for 20 weeks before giving a further 8 weeks of 7-hour days and repeating the step-up programme (Figure 14.1). This regime can be used to achieve 6 laying cycles in 4 years.

Novel lighting programme

It is possible to achieve out-of-season egg production by supplementing natural light with artificial light to make a 20-hour day from about a month before the anticipated normal start of egg production and then transferring them back to natural lighting when they start to moult (after about 3 months lay). The geese will then spontaneously start a second laying cycle about 2 months later (see Figure 3.36, page 46).

Total egg production in a 12-month period will be similar to that for naturally lit geese, but the egg income will be higher in countries where out-of-season hatching eggs and goslings command a premium.

Light intensity

The evidence for the effect of light intensity on egg production in geese is equivocal. Research in Taiwan showed that there is minimal influence on rate of lay between 10 and 220 lux when

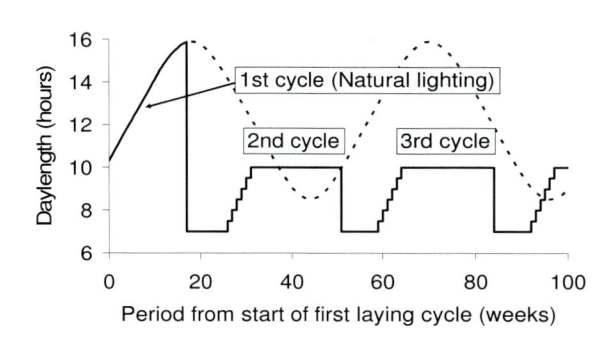

Figure 14.1 *Lighting programme to produce 6 laying cycles in 4 years in geese that have had one cycle under natural lighting. Subsequent cycles are achieved by repeating the 8 weeks of 7-hour days, 6 weeks of 30-minute increases and 20 weeks of 10-hour days. The dotted line represents natural daylengths for Hungary (47°N).*

birds are given a combination of natural and artificial illumination. In contrast, Israeli workers concluded that geese will have better persistency of egg production if natural daylengths are extended to an 18-hour day with artificial light at 20 lux rather than at 50 lux (see Figure 5.16, page 90). Geese in Taiwan naturally breed at a different time of year from geese in Israel, so the optimal light intensity of supplemental artificial lighting might vary from one region to another.

Light colour and source

There is no evidence that the use of special lamps or particular colours can result in any better performance than can be achieved under white light.

Use of a lighting programme for geese

It is obvious that lightproof housing must be available to implement a fully controlled lighting programme for geese. However whether it is economic to invest in such facilities will depend on the relative value of eggs and goslings produced out-of-season.

It seems that whichever type of lighting is provided, geese are unlikely to have a laying cycle much longer than 4 or 5 months, and that the only way to increase the annual egg production is by using programmes to reduce the interval between cycles such as the one illustrated in Figure 14.1.

• Ducks •

As noted before, most of the world's ducks live in Asia under natural daylight conditions where the question of a lighting programme does not arise. Although these birds are usually herded into a night shelter to protect them from predators, the accommodation is not lightproof and, in most cases, no mains electricity is available, so that even extending the winter daylength is out of the question. However, most of the ducks in Asia are of egg-laying breeds and will comfortably produce yields in excess of 250 eggs per bird without supplementary lighting.

• Meat-type parents •

Daylength
Ducks
In more northerly climates and especially with Pekin parent stock bred for meat production, there is a marked effect of season on productivity if only natural daylight is used. Here it is common practice to house ducks throughout their lives, though they may in some cases have access to outdoor runs. It has been shown that if ducks are reared under natural daylight at 52^0N there are seasonal differences in laying performance, even when the birds are all given the same schedule of artificially maintained long days during lay (see Figure 3.38, page 47). These seasonal effects can be overcome and excellent laying performance achieved by giving laying and breeding ducks a 17-hour day throughout rearing and laying (see Figure 3.40, page 48). This cuts out all seasonal effects without requiring lightproof facilities. Some breeders recommend starting the ducks at day-old on a 23-hour day and reducing this to 17 hours over the course of the first week, but there seems to be no evidence that this gives any better results than simply starting the 17-hour day at day-old.

Drakes
Males do not need any special lighting programme to achieve normal sexual maturity before the females come into lay. They can be reared in natural daylight or given the same constant 17-hour days as the ducks.

Light intensity
There is no critical evidence for ducks about threshold light intensity for photoperiodic stimulation. It is likely that the threshold is not more than 10 lux, although houses for breeding and laying ducks are commonly lit to provide 20-30 lux at head height.

Light colour and source
White light is effective in duck houses and there is no evidence that the use of special lamps or particular colours will result in any improvement in performance.

• Duckling •

Daylength
The normal practice with duckling reared intensively for meat production is to provide a 23-hour artificial day to stimulate rapid early growth. This certainly works. The possibility that, as with broilers and Muscovy ducks, intermittent lighting might give the same growth rate with lowered feed intake and reduced mortality has not been tested experimentally. However, evidence from commercial trials of a repeating 1 hour of light and 2 hours of darkness shows that it takes about a day longer to achieve the same body weight at slaughter as birds given a 23-hour daylength but, more importantly, that there is likely to be an increase in the percentage of downgrades due to scratching and bruising, and a reduction in feather yield.

Light intensity
In the case of ducks grown for meat in windowless houses, it is usual to start with bright lighting (10-30 lux) for the first 10 days and then to reduce to 1-5 lux by about 2 weeks of age to control vices.

Light colour and source
As for breeding ducks, there is no evidence that the use of special lamps or particular colours results in any better growing performance than white lighting.

Index

Page numbers in bold refer to main sections

A
Accommodation, 2
Acuity, 2
Adaptation of the eye, 2
Age at first egg *see* Sexual maturity
Ahemeral cycles, 2, **65-70**
    laying hens, 135-136
Ametropia, 103
Anti-rachitic properties of lamps, 114
Asymmetric discharge of lamps, 2, 114
Asymmetrical intermittent lighting, 57-65, 132

B
Barbarie ducks *see* Muscovy ducks
Behaviour,
    effects of illuminance, 86-87
    effects of light source, 116
    effects of wavelength, 12, 98-99
Biomittent lighting, 57, 59-60, 132-134,
Blue light *see Wavelength*
Body checked eggs, 50
Bright:dim ratios and entrainment, 84
Broiler breeders, 17, **32-38, 137-144**
    ahemeral cycles, 69
    illuminance, 85
    intermittent lighting, 61-62
Broilers, **38-40, 145-148**
    continuous darkness, 72
    continuous illumination, 70
    effects of illuminance, 86-87
    effects of light source, 115
    effects of wavelength, 95-96
    eye abnormalities, 104-105
    intermittent lighting, 62-63
Bünning's hypothesis, 14, 15
Buphthalmia, 103-105

C
Candela, 3, 79
Carry-over effect, 15
Circadian oscillators (clock), 3, 9, 14-15
Circannual oscillators, 3
Civil twilight, 3, 49
Clux, 3, 8, 79-80
Colour, *see* Wavelength
Colour coding of lamps, 3
Coloured lamps, 113
Colour rendering index, 3
Colour temperature, 3, 112
Cockerels, 71, 97
Cones in the retina, 3, 7
Continuous darkness, 106-107
Continuous illumination, 70-72, 104-106
Corneal radius, 104
Cornell lighting system, 57, 60, 133-134
Critical fusion frequency, 3, 113

D
Dawn and dusk, 3, 49-50, 130, 144, 147
Day, 3
Daylength *see* Photoperiod
Dimming, 81
Dioptre, 3, 103-104
Dopamine, 9
Drakes, 11, 164
Ducks, 10, **46-48, 164**
    effects of illuminance, 90

E
Egg numbers,
    broiler breeders, 33, 35, 37, 85, 142
    ducks, 46-47
    geese, 45, 90
    layers, 18, 28, 60, 66, 72, 82-83, 98, 122, 133-134
    turkeys, 42-43, 70, 88-89, 116, 151
Egg weight
    broiler breeders, 33
    layers, 28, 32, 67, 83, 98
    turkeys, 43
Energy expenditure, 62-63, 85
Enkephalins, 9
External coincidence model, 14
Eye, 9, 103-107
    not essential for stimulation, 10

F
Feed intake or feed conversion
    broilers, 39, 62, 97, 146
    growing pullets, 24, 27, 82
    laying hens, 16, 29, 60-61, 83, 115-116, 126-127

Feed intake or feed conversion *contd*
    turkeys, 40-41, 71, 155
Fertility, 48, 97, 116, 152-153
Finches, testis responses, 13
Flash lighting, 57
Fluorescence, 3
Fluorescent lamps, 98, 112-115, 144
Flux, 3
Focal length, 3
Follicle stimulating hormone, 9
Foot candle, 3
Free running, 57, 58, 133
French system, intermittent lighting, 60, 134-135
Frequency (Hertz), 3, 114

G
Gallilumens, 80
Gallilux, 3, 8, 79-80
Geese, **45-46, 162-163**
    effects of illuminance, 89-90
Glaucoma, 103
Green light *see* Wavelength,
Growing pullets, **23-27, 121-136**
    ahemeral cycles, 65
    continuous darkness, 72
    effects of illuminance, 81-82
    effects of light source, 115
    intermittent lighting, 59
Growth,
    effects of illuminance, 86, 89
    effects of photoperiod, 38-42
    effects of wavelength, 95-97

H
Heat production, 62, 63
Hue, 3
Hypothalamus, 9-12, 15

I
Illuminance, 4, **79-90,** 107
    broiler breeders, 139, 143-144
    broilers, 147
    ducks, 164
    geese, 163
    growing pullets, 125
    laying hens, 129
    turkeys, 151-153, 157, 159
Incandescence, 4, 111
Intermittent lighting, **57-65**
    broilers, 146-148
    laying hens, 132-135
    turkeys, 158
Internal coincidence model, 14
Inverse-square law, 4, 79
Intensity *see* illuminance
Irradiance, 4, 8, 80

L
Lamp efficiency, 80
Laying hens, **28-32, 125-136**
    ahemeral cycles, 65-69
    continuous darkness, 72
    continuous illumination, 71
    effects of illuminance, 82-85
    effects of light source, 115-116
    intermittent lighting, 59-61
    noise output, 58
Leg disorders, 38-40
Lethargy, 107
Lighting programmes *see* Photoperiod
Light intensity *see* Illuminance
Light leakage, 125, 130
Light meter, 79
Lightproof houses, 121-124, 137, 140-143
Light reception, 9
Light source, **111-118**, 125, 144, 147, 160, 163, 164
Lumen, 4, 80
Luminance, 4
Luminescence, 111
Luminous efficacy (of lamps), 4, 80, 111, 112
Luminous efficiency (of the eye), 2, 8
Luminous flux, 5
Luminous intensity, 5
Luminous power, 5
Luteinizing hormone, 9, 15, 31, 34, 44
Lux, 5, 8, 79

M
Mallard, 11, 46
Melatonin, 6, 9, 15, 58, 107
Midnight feeding, 156-157
Modulation, 5
Mortality, 29, 38-41, 61, 84, 116, 145-146
Muscovy ducks, **44-45, 161-162**

N
Natural light, 27, 38, 41, 46, 47, 130-131, 138-139, 143, 157
Night, 5
Night interruption programmes, 58
Noise output, laying hens, 58

Index

O
Oil droplets in the retina, 7
Ovarian carcinoma, 108, 154
Oviposition time, 30, 33, 48, 68, 71, 73, 115, 128, 142

P
Pelleted feed, 23
Phase reversal, 5, 14
Phosphors, 5
Photoinducible phase, 5, 13-14, 57
Photon flux, 5
Photoperiod, 5, **23-50, 57-73**
    breeding turkeys, 149-153
    broiler breeders, 137-144
    broilers, 145-146
    ducks, 164
    geese, 162-163
    growing pullets, 121-125
    growing turkeys, 155-158
    laying hens, 125-136
    muscovy (barbarie), 161-162
Photoperiodic time measurement, 14-15
Photoreception, 9
Photorefractoriness, 16-18
    lack of, in hens, 18
Photopic vision, 2, 5, 7, 8, 79
Pigeon, 10
Pineal reception, 9
Planck's constant, 1, 5
Polycystic ovarian follicles, 106, 154
Poultry *see* Broilers, Ducks, Geese, Laying hens, Turkeys etc.
Prolactin, 16
Pullets *see* Laying hens

Q
Quail, 10, 13
Quantum energy, 5

R
Reading system of lighting, 57, 60-61, 133-135
Radiance, 5
Radiant flux, 5
Recycling turkey hens, 87-88, 153
Red light *see* Wavelength
Red mite control, 134
Reflectance, 80
Refractive error, 104
Retinal reception, 9, 103
Rods in the retina, 5, 7

S
Saw tooth light programme, 16
Scotoperiod, 6
Scotopic vision, 6, 7
Seasonal effects *see* Natural days
Sexual maturity
    broiler breeders, 17, 18, 32, 35-37, 138-141
    effects of illuminance, 81
    effects of wavelength, 97-98
    geese, 45
    pullets, 18, 23-27, 72, 81, 121-126
    turkeys, 43, 70, 88, 116
Shell quality, 28, 34, 67
Short cycle intermittent lighting, 57, 60
Sodium vapour lamps, 113
Sparrow, 10
Spectral luminous efficiency, 6
Spectral sensitivity, 7
Speed of light, 6
Steradian, 6, 79-80
Step down lighting, 24, 124
Step up lighting, 128, 142-3, 157
Stress, 99, 105-106, 108
Subjective day, 6
Sunlight, 99
Suprachiasmatic nucleus, 9
Symmetrical intermittent lighting, 58, 132-135

T
Testis weight, 11, 13, 89, 97
Thyroxine, 10
Time of lay *see* Oviposition time
Turkeys, **40-43, 149-160**
    ahemeral cycles, 70
    continuous illumination, 71
    effects of illuminance, 87-89
    effects of wavelength, 95
    eye abnormalities, 105
    intermittent lighting, 63-65
    photorefractoriness, 17-18
Twilight *see* Dawn and dusk

U
Ultraviolet light, 6
    UV-A, 6, 99, 108, 112, 118, 144, 159
    UV-B, 6
    UV-C, 6
Unconventional photoperiods, **57-73**

V
Velocity of light, 6

Vision, 7-8
    poultry compared to humans, 2, 7

W
Waterfowl *see* Ducks, Geese, etc.
Wavelength, 81, **95-100, 111-118**
    broiler breeders, 144
    broilers, 147
    human colour sensations, 1
    layers, 130-131
    pathological effects, 107
    pullets, 125
    stress, 108
    transmission through skull, 10-12
    turkeys, 153, 159-160
Welfare, 118, 129

Z
Zeitgeber, 6, 14